The Complete Hiker's Guide To
# The Backbone Trail

Doug & Caroline Chamberlin

Published by Riviera Books
1125 E. Broadway
Suite 702
Glendale, CA 91205

*Although the authors strive to give adequate warning about the hazards along any particular hike, situations may change which are beyond the scope of this book to anticipate. Fires can alter vegetation, landslides can change the terrain, footpaths can become overgrown, ruins might have become unstable, parking restrictions may be changed and authorities may occasionally close trails. No hiking guide is a substitute for your own judgment. If you feel unsafe or unsure about a particular stretch of trail, play it safe and exercise caution or turn back. The authors and publisher take no responsibility for a trail or area which has become hazardous or is unsuitable for certain visitors due to their age, physical ability, imprudence or lack of preparation. They are also not responsible for any damage, injury or loss occurring while using this book.*

*While the authors strive to make this book as accurate as possible, they make no claims, promises, or guarantees about the accuracy, completeness, or adequacy of its contents, and disclaim liability for any errors and omissions contained herein.*

*Reference in this book to any company, organization, event, product or service is for information only and does not constitute endorsement or recommendation by the authors or Riviera Books. Readers are expected to do their own due diligence.*

*This book is dedicated to the visionaries
who made the Backbone Trail possible,
to the volunteers who keep it passable,
and to the activists who keep it accessible.*

# PREFACE

Necessity, they say, is the mother of invention. If that's true, then the mother of this book was a New Year's resolution. As the ball dropped at midnight on December 31st of 2010, your co-author Doug Chamberlin announced to his wife Caroline that in the new year he would get in shape. Big deal. "This year I will get in shape" is the most clichéd and least kept New Year's resolution in the history of well-intentioned fibs.

Knowing fully well how lame that sounded, we both decided to figure out a way to make getting exercise fun. So instead of buying another wasted gym membership, we announced boldly that in the new year we would (A) hike the entire Backbone Trail and (B) climb Mount Baldy. Baldy turned out to be the easier feat. The Backbone Trail proved to be a challenge in a way we hadn't anticipated. There was no complete hiking guide available. Of course, we could buy maps, and several excellent hiking books covered certain parts of the Backbone Trail, but those were not particularly helpful when we found ourselves at junction after junction with no signage and a map that was too small to guide us. Getting lost was actually fun at times – for a few minutes – and then it quickly got old. Doug began to complain, "Someone ought to write a complete guide to this thing!" That's when he got hit with the ol' "someone oughta." That phrase is frequently thrown back at Doug by his spouse whenever he complains that "someone oughta" fix something that is not right with the world. She always claims, "That someone should be you." As usual, she was right.

So, we set about creating a hiking guide to the trail. Backpacking it was not really an option at the time, mainly because there were few places to camp along the trail. And yet we wanted to hike the trail through. So we hiked it, start-to-finish, in day-long segments. Each segment was defined by wherever a public road crossed the trail.

Since then, we have hiked the trail multiple times, both eastbound and westbound. Many of the details referred to in this book are from personal experience (with the exception, of course, of history, basic facts and quoted references). So, when you read, "This part of the trail is particularly steep and gravelly; use caution to avoid falls," you wouldn't be off base to picture one of your authors slipping and falling right on their butt and into a cactus at that very spot. If we mention the possibility of spotting wildlife, it may be because it's the place where Doug rounded a curve and spotted a mountain lion for a few thrilling moments.

This book is written with the average hiker in mind. The reasons are 1) most readers of this book will be in the "average" hiking range and 2) we ourselves are average hikers. We are not hard-core hiking types with humongous backpacks, and we're also not total beginners out for a tentative walk in the woods. We are right in the middle – two average jamokes who enjoy hiking around Southern California – and this book is written with that expectation in mind. (If you are hard-core or a total beginner, simply adjust your expectations upwards or downwards per the book's description of hiking difficulty.)

The point of all this is that if we can do it, you can too. And you should. On this wonderful trail you'll find a badly needed escape from the so-called "real world." The more time you spend on the Backbone Trail, the more your definition of "the real world" will shift – and for the better, we think. As John Muir wrote in 1901, "Thousands of tired, nerve-shaken, over-civilized people are beginning to find out that going to the mountains is going home; that wildness is a necessity; and that mountain parks and reservations are useful not only as fountains of timber and irrigating rivers, but as fountains of life."

Tired? Nerve-shaken? Over-civilized? Welcome to the club. Let's explore this wonderful trail together. On it you'll find peace, beauty, inspiration, health and adventure.

But first, please buy the book!

Doug & Caroline Chamberlin

# ACKNOWLEDGMENTS

We wish a special and heartfelt "thank you" to the following wonderful people who helped in the creation of this book, and to whom we are particularly indebted:

To Melanie Beck of the National Park Service, Sheila Braden at the Anthony C. Beilenson Interagency Visitor Center, and all the staff and volunteers at the Center for providing invaluable information and fielding our sometimes goofy questions.

To Sophia Wong at Western National Parks Association, David Haake of the Sierra Club West L.A. Group, Joe Phillips of the Sierra Club San Fernando Valley Group, and John Mead of Adventure 16 for both championing this book and for fostering some very fun Backbone Trail presentations.

To the truly inspiring folks at the Santa Monica Mountains Trails Council, in particular Ruth Gerson, Jerry Mitcham, Sheryl Phelps and Howard Cohen for their advice, information and support. To Francine Marlenee, Tyson Gaskill and Vicki Rea-Gaskill for their helpful editorial advice, Melinda Augustina for indispensable marketing guidance, and Nancy Pearlman of Environmental Directions Radio for helping get the word out about the book.

To hiking buddies Lisa Hart and Ethan Barrett, as well as Tammy Caplan and Joe Gold, *fans extraordinaire* of this project. Also, to Herb and Deena Chamberlin, who long ago encouraged a love of walking in the mountains.

And finally, thanks to all the Backbone enthusiasts who purchased the first edition and wrote us with comments, questions and suggestions, many of which we have incorporated into this edition.

# CONTENTS

**Introduction**

Chapter 1: Welcome To the Backbone ...................................................... 10
  *What's So Great About The Backbone Trail?*
  *Why A Backbone Trail?*
  *A Brief History Of The Backbone Trail*
  *Just How Long Is The Backbone Trail?*
  *The Springs And Woolsey Fires*

Chapter 2: How To Use This Book .......................................................... 18
  *Why Use A Backbone Trail Guidebook?*
  *How Each Chapter Works (Including Map Legend)*
  *Through-Hiking The Backbone Trail (And Alternatives)*
  *A Brief Glossary*

Chapter 3: On The Trail.......................................................................... 27
  *General Hiking Tips*
  *Hike Hacks*
  *Trail Hazards*
  *Trail Etiquette*

Chapter 4: Flora, Fauna & Faultlines.................................................... 34
  *Geography*
  *Geology*
  *Plants*
  *Animals*

**The Hikes**

Segment 1: Will Rogers Park to The Hub ............................................... 44
  *("Rogers Trail")*

Segment 2: The Hub to Trippet Ranch .................................................. 60
  *("Musch Trail" & "Eagle Rock Fire Road")*

Segment 3: Trippet Ranch to Topanga Canyon ..................................... 68
  *("Dead Horse Trail")*

Segment 4: Topanga Canyon to Old Topanga Canyon ......................... 74
  *("Henry Ridge Crossing")*

Segment 5: Old Topanga Canyon to Saddle Peak Road....................... 78
  *("Hondo Canyon Trail")*

Segment 6: Saddle Peak Road to Lois Ewen Overlook.......................... 85
  *("Fossil Ridge Trail")*

**Segment 7: Lois Ewen Overlook to Saddle Peak** ........................................ 90
 *("Saddle Peak Trail" East Approach)*

**Segment 8: Saddle Peak to Stunt Road Cutoff** ............................................ 95
 *("Saddle Peak Trail" West Approach)*

**Segment 9: Stunt Road Cutoff to Piuma Road** .......................................... 101
 *("Saddle Creek Trail")*

**Segment 10: Piuma Road to Las Virgenes Road** ...................................... 108
 *("Piuma Ridge Trail")*

**Segment 11: Las Virgenes Road to Corral Canyon Road** ..................... 114
 *("Mesa Peak Motorway")*

**Segment 12: Corral Canyon Road to Latigo Canyon Road** ................. 123
 *("Castro Crest")*

**Segment 13: Latigo Canyon Road to Kanan Dume Road** ..................... 130
 *("Newton Canyon Trail")*

**Segment 14: Kanan Dume Road to Zuma Ridge Motorway** ............... 135
 *("Upper Zuma Canyon Trail")*

**Segment 15: Zuma Ridge Motorway to Encinal Canyon Road** .......... 141
 *("Trancas Canyon")*

**Segment 16: Encinal Canyon Road to Mulholland Highway** ............ 148
 *("Clark Ranch Trail")*

**Segment 17: Mulholland Highway to Yerba Buena Road** .................. 153
 *("Etz Meloy Motorway")*

**Segment 18: Yerba Buena Road to Triunfo Pass** .................................. 161
 *("Mount Triunfo")*

**Segment 19: Triunfo Pass to Big Sycamore Canyon** ........................... 169
 *(Via Boney Mountain)*

**Segment 20: Big Sycamore Canyon to La Jolla Canyon** ..................... 190
 *("Ray Miller, Overlook & Wood Canyon Vista Trails")*

## Appendices

Quick-Reference Charts ........................................................................... 200

Index ......................................................................................................... 204

About The Authors ................................................................................ 211

# THE BACKBONE TRAIL

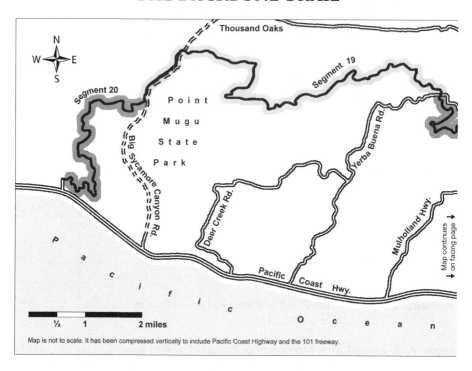

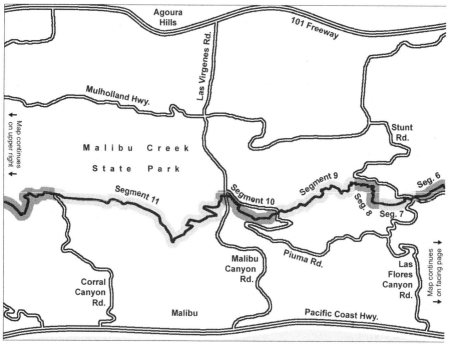

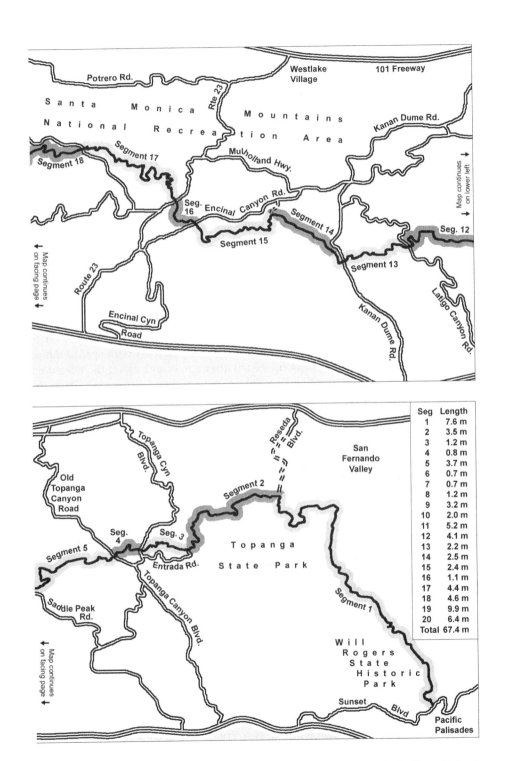

Map labels (top map):

Potrero Rd.

Westlake Village

101 Freeway

S a n t a  M o n i c a  M o u n t a i n s
N a t i o n a l  R e c r e a t i o n  A r e a

Rte 23

Kanan Dume Rd.

Segment 17

Segment 18

Mulholland Hwy.

Seg. 16  Encinal  Canyon  Rd.

Segment 14

Segment 15

Seg. 12

Segment 13

Route 23

Encinal Cyn Road

Kanan Dume Rd.

Latigo Canyon Rd.

Map continues on lower left →

← Map continues on facing page

Map labels (bottom map):

Topanga Cyn Blvd.

Reseda Blvd.

San Fernando Valley

Old Topanga Canyon Road

Segment 2

Seg. 3

Seg. 4

Segment 5

Entrada Rd.

T o p a n g a  S t a t e  P a r k

Saddle Peak Rd.

Topanga Canyon Blvd.

Segment 1

W i l l  R o g e r s  S t a t e  H i s t o r i c  P a r k

Sunset Blvd

Pacific Palisades

← Map continues on facing page

| Seg | Length |
| --- | --- |
| 1 | 7.6 m |
| 2 | 3.5 m |
| 3 | 1.2 m |
| 4 | 0.8 m |
| 5 | 3.7 m |
| 6 | 0.7 m |
| 7 | 0.7 m |
| 8 | 1.2 m |
| 9 | 3.2 m |
| 10 | 2.0 m |
| 11 | 5.2 m |
| 12 | 4.1 m |
| 13 | 2.2 m |
| 14 | 2.5 m |
| 15 | 2.4 m |
| 16 | 1.1 m |
| 17 | 4.4 m |
| 18 | 4.6 m |
| 19 | 9.9 m |
| 20 | 6.4 m |
| Total | 67.4 m |

# CHAPTER 1:
# WELCOME TO THE BACKBONE

When was the last time you had an adventure? Not a huge adventure, one full of peril and exotic wonder, but a pocket-sized adventure. One you can have practically in your own backyard and in your spare time. You can have this on the Backbone Trail – nearly 70 miles of *local* adventure.

If you live in Southern California, you may already think of yourself as lucky. But if you love the outdoors, your luck is greater than you estimated because you live near the Backbone Trail, one of the best-kept secrets in SoCal. Most people don't have a gem like this anywhere near the outskirts of their city. Most hikers would beg to have a trail like this so close to home.

The Backbone Trail, the 67.4-mile-long hiking path which follows the spine of the Santa Monica Mountain Range from Los Angeles to Point Mugu in Ventura County, is under-appreciated. Many hikers have checked out small segments of it on a single day-hike, only to wonder where the rest of it goes. But where it goes isn't really the point, in the same sense that the reason for any great trip isn't to get from point A to point B. What matters more is what you'll experience while exploring the Backbone. Sooner or later – and it won't take long – something along this trail will work its way into your soul, and you'll always remember it.

It may be as you stand atop a windy peak with miles of ocean spread out at your feet, or when you emerge at the beach after a long trek along the mountain spine. It may be your view of endless ridges as they disappear into the haze, or watching the city lights come on as the sun sets into the pale sea. It may be the babbling stream you never expected, or moss-covered boulders in a gloomy bay tree forest, or ducking through a tree tunnel of ceanothus.

It may happen as you cross a hillside green with brilliant clover, or discover mushrooms hidden in a shady nook. It may be when you first spy mountain lion tracks along the path, or hear beautiful birdsong in an elfin forest. Or when you round a bend and spot a hillside of blooming white yucca, or walk a path lined with thousands of orange monkey flower blossoms. It may be ice on the trail on a brisk winter day, or a cool drink of water from a fire road spigot on a warm spring afternoon. Or it may simply be the moment when you look around and realize, "I'm the only person out here!"

All of this and much more awaits you on the Backbone Trail, so needless to say, we think YOU should hike it. And not just part of it. Why not hike all of it? Maybe not all in one stretch with a backpack, but the way we've done it several times – by day-hiking it in segments. Although through-hiking the trail is currently challenging due to inadequate camping areas, it makes for a wonderful series of day trips that add up to more than just random day hikes. By doing the Backbone segment-by-segment, you'll get a sense of the span of the Santa Monica Mountain Range from end to end. You'll experience its slow transition from the rolling meadows and oak-covered slopes of the east to the dramatic, craggy, volcanic drop-offs of the west. You'll notice as the semiurban oak parkland in the east dissolves into the austere wilderness of arid mesas at the trail's western end.

*Sunrise along the Backbone Trail above Triunfo Pass*

And with you all the way will be the constant companion of the rolling chaparral, the endless spine of rocky peaks, and the majestic Pacific Ocean laid out at your feet like a royal blue carpet.

## WHAT'S SO GREAT ABOUT THE BACKBONE TRAIL?

The Backbone Trail, an officially designated National Recreation Trail, holds a unique place amongst America's great hikes and is well deserving of its own book. But why? What's so special about this footpath? Plenty of trails traverse much more spectacular mountains, many boast more beautiful forests, and some (like the Appalachian and Pacific Crest Trails) are far longer. So why does this thing merit special attention?

For one, there's the Pacific. Although many long trails offer lovely forests, peaks, waterfalls, and the occasional deer spotting, how many of these also offer nearly constant views looking down from hundreds of feet on the Pacific Ocean? Try *that* on the Appalachian Trail. By contrast, a Backbone Trail daytrip on which you hike through deep woods, cross babbling streams, climb rocky crags to gaze out over endless mountain ranges, and then turn around to look down upon a thousand square miles of ocean is fairly routine.

Secondly, the Backbone Trail has the distinction of being a long trail at the back door of the second largest metropolis in the United States. Although you may feel at times while on the trail that you are a hundred miles from the middle of nowhere, you're actually a hundred miles from the middle of 18.5 million people! In fact, it's estimated that 1 in 17 Americans lives within an hour's drive of the Backbone Trail. Topanga State Park, through which part of the trail runs, is considered the world's largest wildland contained within the boundaries of a major city. All of this makes the Backbone potentially one of the most popular recreational trails in the world.

Third, the Backbone's east-west orientation is rare for long trails. This is because the Santa Monica Mountains, along whose "backbone" it runs, is a transverse range (meaning it runs east-west instead of north-south like nearly all major American ranges). This may not matter to you when

*A quiet pool in Dark Canyon*

you're out along the trail, but geographically it makes the thing a bit of a star.

Fourth is the real estate through which the trail runs. As they say in the biz, it's all about location, location, location. And when you're on the Backbone Trail, your location is... wait for it... Malibu! Yep, out here it's all movie stars, media moguls and you. Okay, maybe this is an exaggeration, but the trail's land donors do include Arnold Schwarzenegger and James Cameron. This is some of the most coveted real estate in the nation, and when you're not enjoying the wilderness, you are passing near the backyards of some very impressive estates. This means that you, in your dirt-covered hiking pants and sweat-drenched floppy hat, have the Richie Rich distinction of hiking through the most expensive national park in the country. With its acquired parcels valued at about $100 million, the Santa Monica Mountains National Recreation Area was created at a higher cost than any other parkland.

Finally, there's the weather. If you live in California, you already know the appeal of our temperate climate, and this extends to the Backbone Trail as well. While most trails are limited to a half-year hiking season, our Backbone Trail offers year-round appeal. Spring brings blossoms of every color, emerald green hillsides, seasonal waterfalls and perfect weather. Winter maintains a cool but pleasant temperature, with occasional icy patches on shaded northern slopes. Fall brings just a hint of leaf color to the sycamores and vines. And then there's the summer – ironically the worst hiking season for the Backbone Trail. Too much sun and heat can turn a hike on a fire road into what some hikers call a "death march" – a miserable trudge that makes you just want to go jump in the ocean. But even in summer, under cloudy or cool conditions, a short hike can be a fully worthwhile endeavor, especially in the early morning or at sunset along a moonlit fire road. When you explore the Backbone Trail, you'll find that the common misconception of it being all sun-exposed chaparral and brown sunbaked hillsides couldn't be more off.

## WHY A BACKBONE TRAIL?

Superlatives aside, why even have a Backbone Trail at all? The trail didn't just make itself. Many people labored, both physically and politically, to create it. Why all the work? Weren't there plenty of hikes available in these mountains already?

There's more than one argument for the creation of a Backbone Trail, but one stands above all the rest. In a word: preservation. The great appeal of this mountain range is also its greatest threat in the form of private development. Every year, more and more parcels in the vicinity are sprouting housing and other "improvements." The main way to stop further despoliation of this beautiful region is to create parkland and preservations out of it. Over the years, increasing chunks of the area have been placed under the protection of the Santa Monica Mountains National Recreation Area, the California State Parks system, and the Santa Monica Mountains Conservancy. But these groups can't or won't continue to preserve more lands without the support of the public. And John Q. Public doesn't tend to fight very hard to preserve areas to which he has no access. That's why you need a trail: so the public can go there, enjoy it, and feel like they have a personal stake in preserving it. If people are using it and appreciate it, they'll object to its despoliation.

This explains the need for trails in general, but why a single Backbone Trail? Because a single trail, tying the entire area together (at least in our minds) tends to create awareness of the Santa Monica Mountain Range as a single ecological and recreational area, giving the argument for preservation a bit more power. With one long trail unifying it, what looks on a map like a hodgepodge patchwork of different kinds of parks instead feels like one large recreational unit. Once the public gets to thinking of it in this way, then harming a small part of it seems like more of a threat to the whole.

*City lights shine during a moonlight hike on Etz Meloy Motorway*

There are other less political reasons for a Backbone Trail. For one, the people of Southern California deserve a long trail for their enjoyment. In addition to the local portion of the Pacific Crest Trail, the 53-mile Silver Moccasin Trail and the proposed Rim-Of-The-Valley Trail, the Backbone Trail provides a lengthier adventure than a self-contained day hike.

The final reason is a bit more enigmatic. There's something in people that wants to unite things together, a certain completist satisfaction in covering something from start to finish. The Blue Ridge has its parkway, the Appalachians have their trail... it seems to somehow legitimize the whole thing. The dream of a single footpath spanning the entire length of the Santa Monica Mountains has been around for decades, and the Backbone Trail may be our attempt to tie a big ribbon around the whole shebang and say, "There, that's packaged nicely." This mysteriously satisfying sense of completism is, for us at least, the appeal of hiking the entire Backbone Trail, and the reason for this book.

## A BRIEF HISTORY OF THE BACKBONE TRAIL

So how do you create a trail along the spine of a rugged mountain range covered extensively by private property and right next to a huge metropolis? The answer is: slowly but steadily. The dream of creating a trail spanning the Santa Monica Mountain Range has existed since before the 1970's, but with the conservation movement of that era, it began to sprout from dream to reality.

According to the Santa Monica Mountains Trails Council, it was Darrell Readmond who first had the foresight to plan a trail (originally called "The Ridge Trail") along the crest of these mountains, and it was Joanne Hubbard who named it "The Backbone Trail." The original plan was for the trail to start in Griffith Park and continue all the way to Point Mugu, making it truly link the entire range from start to finish. But with rapid development in the Hollywood Hills, it was clear that a trail through that section would be next to impossible. (That dream is still alive. The creation of the Betty Dearing Mountain Trail in Studio City adds a charming section of footpath to the eastern range that could in theory connect the Backbone Trail to the trails of Griffith Park.) At any rate, the trail plan was re-envisioned to start at Will Rogers State Historic Park in Pacific Palisades.

The California State Legislature approved funding to create a Backbone Trail in 1974. With the establishment of Malibu Creek State Park that same year, a decent chunk of the trail's real estate was now secured by four state parks.

But about 34 miles of the proposed trail still crossed private land, so in 1978, after years of lobbying and activism from environmental advocates like Susan Nelson, Margot Feuer and Jill Swift, congress created the Santa Monica Mountains National Recreation Area to begin filling in the blanks. Piece by piece, the National Recreation Area began acquiring parcels to create federal parkland.

By 1980 another entity was acquiring parcels: the Santa Monica Mountains Conservancy. Established by the state legislature, its mission is to "strategically buy back, preserve, protect, restore, and enhance treasured pieces of Southern

*A quiet afternoon on the Eagle Springs Fire Road (alternate route)*

California to form an interlinking system of urban, rural and river parks, open space, trails, and wildlife habitats that are easily accessible to the general public."

These three landholding entities worked together, along with the Sierra Club, the Santa Monica Mountains Trails Council, and other volunteer organizations to secure the land on which the Backbone Trail now lies. The preservation process has slowed in recent years due to skyrocketing real estate values and slashed government budgets, but it continues forward nonetheless.

In the meantime, a group of energetic volunteers was planning and building the trail itself. Large parts of the proposed Backbone route already existed in the form of older trails and about twenty miles of ranch roads and fire roads. The rest of the route had to be built by hand. A handful of hardy volunteers began "hiking" the proposed route... but without a trail to walk on. This involved crawling and bushwhacking through thick chaparral while planting flags to mark the route.

In 1980, the actual trail-building began with the Dead Horse segment. Within a few more years, some of the best parts of the Backbone Trail – Newton Canyon, Saddle Creek, The Musch and Ray Miller Trails – were complete, and by 1999 forty uninterrupted miles of trail between Will Rogers State Historic Park and Zuma Ridge Motorway were open. More sections to the west opened in the following decade, built to a gentler grade using heavy equipment. Finally, in June of 2016, the National Park Service acquired the final parcels of land to complete the trail.

But who *really* built the thing? Most of the physical trail construction was done by volunteers – many of them from the Santa Monica Mountains Task Force Trail

Crew, a part of the Angeles Chapter of The Sierra Club. This ever-changing group of heroes built or helped build the Musch, Dead Horse, Piuma Ridge, Ray Miller, Saddle Creek, Fossil Ridge, Hondo Canyon and Wood Canyon Vista segments.

Another organization that was essential in constructing the trail is the Santa Monica Mountains Trails Council. Formed in 1969 by a group of equestrians who were concerned about encroaching development, the council soon grew to include hikers, runners, and bikers. Their mission is to preserve, create, and maintain trails in these mountains. A trail doesn't just maintain itself, and if you've ever enjoyed a hike during which you didn't have to push your way through overgrown tangle or hop across washouts, you might want to thank these passionate volunteers.

Much of the building was also done by the California Conservation Corps, a state agency that encourages young men and women to sign up for a year of working outdoors to improve California's natural resources. Their slogan is "Hard work, low pay, miserable conditions... and more!" Hard hats off to these folks.

The National Park Service also built portions, like the Newton Canyon and Castro Crest segments, in addition to overseeing the process. Park rangers, maintenance staff and engineers all played an important role.

Taking into account all the hardworking folks involved, people always want to know is there a "Father" or "Mother of the Backbone Trail?" Should any one person get the top credit, in the way that Benton MacKaye is often credited for the Appalachian Trail? The answer is no, as the individuals responsible for the creation of this gem are too numerous to mention, but there are at least two names who still deserve a shout-out. They are Ron Webster and Milt McAuley.

Milt McAuley was an avid hiker, nature lover and Sierra Club outing leader who promoted the idea of a Backbone Trail as early as anyone. He wrote hiking guides, including one (now out of print) promoting the Backbone Trail in its infancy, which introduced many to the Santa Monica Mountains. His work helped spread the word about the Backbone Trail as not just an idea, but a concrete thing that was going to happen. He was a beloved figure within the Sierra Club to such a degree that upon his passing in 2008, an effort was begun to rename "Peak 2049" along the Backbone Trail as "McAuley Peak." It was officially renamed in 2015.

Ron Webster may be more responsible for the physical lay of the trail than any other person. Ron, who passed away in 2021, had a unique job: professional trail builder. He started off building trails as a volunteer, then became the leader of the Santa Monica Mountains Task Force Trail Crew at about the time that the Backbone project was really heating up. With a grant from the Santa Monica Mountains Conservancy, he planned and oversaw construction on vast quantities of the Backbone Trail. In a 1994 L.A. Times interview, he explained, "I try to build them so that people don't think they're walking on a trail. I want them to think they're just walking in the woods." He added, "There's a lot of trail manuals and schools... I didn't go to any of them. But I remember something that lots of trail builders forget: The trail should lay lightly on the land."[A] Anyone who has ever hiked the breathtaking Hondo Canyon section – a masterpiece of trail design – will witness his work speak for itself.

---

A. Muir, Frederick. (1994, December 28th). Lightly on the Land. *Los Angeles Times*.

## Just How Long Is The Backbone Trail?

The correct answer is: "It depends on who you ask and when you ask them."

We're not being snide. It's true. Different authorities don't fully agree on just how long the trail is. They all come in relatively close, but if you consult the National Park Service, the U.S. Geological Survey, Google Maps, hiking websites, guidebooks and trailhead signs, you'll get a slightly different answer from each.

What gives? Is the Backbone Trail flexible like a rubber band? Isn't distance supposed to be pretty much a stable, reliable thing? Well, yes and no. It turns out that trail measuring is actually a bit of an inexact science. Try it on a hike with your own GPS device and you'll get a slightly different result each time.

To further complicate it, the National Park Service's re-measurement of the trail after its 2016 completion came in a little shorter than their original 2008 numbers. The first edition of this book used the 2008 length (adjusted for some road bypasses of private property). Since the NPS updated their numbers, we've updated ours for this edition. So, as far as we're concerned, the Backbone Trail is – wait for it – 67.4 miles. That's our story and we're sticking to it. At least until the NPS re-measures it.

## The Springs And Woolsey Fires

In a span of five years, the Santa Monica Mountains have been hit by two major fires. In 2013, the Springs Fire started along the 101 freeway and raged south through Point Mugu State Park all the way to the Pacific, affecting the westernmost nine miles of trail. Years of subsequent drought inhibited the return of vegetation. Then, in 2018, the infamous Woolsey Fire burned nearly 100,000 acres, including 88% of the National Recreation Area and most of Malibu Creek State Park. The only thing that stopped its westward march was the footprint of the Springs Fire. Together, these two blazes impacted about 41 nearly continuous miles of the trail.

Unfortunately, huge *megafires* seem to be the way of the future. Although this region is used to fire and has evolved to expect and rely on it, that's not the case with frequent megafires. Too much burning prevents the native vegetation from getting a foothold, leaving a more arid landscape to become the new normal.

In spite of this, hiking in the burn areas can still be pleasant. Most of the damage from the Springs Fire is fading into the past. In the Woolsey Fire area, chaparral is returning to the hillsides, although the charred skeletons of old shrubs are still present. Many of the old oaks, tough as nails, stand with fully blackened trunks, sporting new foliage nonetheless. The sycamores didn't fare as well, but new ones will sprout from old roots and can grow up to two feet in a good year.

In a few more years – barring another conflagration – the chaparral will look like it did before, since many plants will re-germinate in the same spots, recreating the former scene. The full-size trees will take longer. For the time being, expect the western half of the Backbone Trail to be in a state of flux. Shade is diminished. Infrastructure is still being rebuilt. Vegetation is changing. Consequently, the descriptions in this book might not fully describe what you encounter.

CHAPTER 2:

# HOW TO USE THIS BOOK

When you first decide to hike the Backbone Trail, you might picture yourself strapping on a backpack and tackling the whole thing, starting at one end and emerging at the other end a week later. Unfortunately, this isn't very practical for the time being. You *could* do it, and some very determined souls do, but the lack of camping along the trail makes this very challenging. (More about how to through-hike it follows).

Instead, this book is mainly intended for day-hikers. You can easily hike the entire trail from start to finish as a series of day trips and still feel like you've really gotten out in the country.

We've broken up the Backbone Trail into twenty segments, each described in its own chapter. Some segments are long, others are short and could be combined with an adjacent segment. Each segment is defined by car access points. Whenever a public road crosses the Backbone Trail, that starts off another segment. Nearly all of these crossings feature parking of some kind or another, and some have fully developed trailheads with restrooms and other amenities.

For the longest stretches, we've cheated a bit by bisecting each into two segments. This is done by accessing the long stretch in its middle via a "shortcut" on an adjoining trail or dirt road. The first and last segments are such examples. Hardier hikers might want to do both segments as one and disregard the shortcut.

We've arranged the hikes linearly from east to west. Segment 1 is the most easterly section, with Segment 2 adjoining to the west, etc. The last segment, Segment 20, is the farthest west.

Why east-to-west? The National Park Service describes the trail from west to east, so why differ with them? There are several reasons. First, heading westbound makes for a more dramatic experience. The scenery in the Santa Monica Mountains tends to intensify as you head west, with more cliffs, rock domes, deeper canyons and higher summits in the western half. There's also more city influence in the east, but by the time you get into the western half, you'll feel like you've trekked out into the middle of nowhere. So which trip would you rather take: one that slowly took you from the rugged wilds back into the city, or one on which you slowly left your city environment behind for something more exotic? Easy choice for us.

Going east-to-west also takes you from generally easier hikes into more rugged ones, which allows you to build up endurance and makes it a little easier on the legs. If you're actually through-hiking the trail, east to-west is easier since the single worst elevation gain, the 3,000-foot climb up Boney Mountain, is changed to a downhill trek.

Another reason to hike east-to-west is that, since most people hike in the morning to avoid the afternoon heat, it's best to have the sun at your back in the morning instead of in your eyes.

All of this having been said, since this book is oriented toward day-hikes, we had to make a difficult decision: should we stay true to the east-to-west orientation of the chapters and describe *every* hike from east to west, regardless of which direction makes a better day-hiking experience? The answer is no. If you're day-hiking, what you really want is the best hike, not some abstract adherence to an east-to-west rule. So we've chosen to describe each segment in whichever direction makes the most sense for that hike. While the majority of trips are described as east-to-west, some are described west-to-east. It depends on pragmatic factors such as:

- **Parking.** Some segments have decent parking only at one end. Naturally, it makes sense to start where there are better parking options.
- **Views.** In some cases, hiking in one particular direction provides better views or increasingly dramatic vistas.
- **Elevation change.** A few long, one-way segments have been oriented so that the majority of the hike is downhill, not uphill.
- **"Destination" hikes.** A few segments lead to a peak or other goal which makes for an obvious turnaround spot. Since you can't start your hike at a peak (unless you own a helicopter), the access must be from the other end of the segment.

All of our mileages were recorded using GPS tracking, then synched to match the mileage numbers published by the National Park Service in 2016.

## WHY USE A BACKBONE TRAIL GUIDEBOOK?

Is this book necessary? Can't you just head out on the trail and see where it takes you? Maybe just look up a brief description online and take off? You could, and that might work for you, but if you'd like to do it right and not get pretty badly lost, you should use a guidebook. That's because, in spite of its many virtues, the Backbone Trail does tend to get confusing at times. It was created from a combination of old paths, dirt roads and newly designed trails, and because of this, there are many junctions, twists, turns and other surprises. There are no blazes. Some junctions are unsigned. Side paths join the main route only to branch off again. Trails turn to dirt roads and back to trails again, and sometimes a single-track path may jog briefly left

*Okay… which one of these things is the Backbone Trail?*

or right along a fire road. None of it is very intuitive. There are plenty of mistakes to be made. We know, because we have managed to make all of them for you! For your convenience, we've described every junction and noted it with GPS coordinates so you can verify it on your mobile device.

You'll also get other details in this book that you shouldn't expect from the web: history, geology, vegetation, as well as accurate mileage, elevation data and maps.

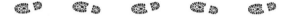

## How Each Chapter Works

Each segment gets its own chapter, regardless of its length. Most of the hikes can be hiked out-and-back (or, with the use of an alternate route, as a loop). The longest segments are intended to be one-way, requiring cars parked at both ends or a car shuttle. Intermediate length segments can be done one-way or as an out-and-back hike. In these cases, we'll give driving directions to both trailheads.

How fast do you hike? Are you a tiger, leaving everyone else in the dust? Are you a flower-sniffer, out for a leisurely stroll with frequent stops? Or are you Joe Average, right in the middle? Since this book is written with Joe Average in mind, this means that when we describe a climb as "tough" or "long," it will be a significant but doable workout for the typical hiker out for an average day-hike. If you're a tiger or a flower-sniffer, adjust your expectations accordingly. If we describe a stretch as taking "ten minutes," add or subtract a few minutes per your personal hiking speed.

The latter won't be much of a problem, since we describe nearly all distances in terms of miles and not in time. The average hiker moves between 2 and 2½ miles an hour. By this rule, a ½-mile stretch should take approximately 12 to 15 minutes.

At the head of each chapter, you'll see the **segment number** and the **segment span** which it covers (for example "Stunt Road Cutoff to Piuma Road.") You'll also see another subtitled name, like "Saddle Creek Trail." This is the original name for that part of the

| Segment 9: |
| --- |
| **Stunt Road Cutoff to Piuma Road** |
| ("Saddle Creek Trail") |
| BBT segment length: 3.2 miles<br>Day hike length: 3.4 miles one-way<br>Suggested day hike: east to west |
| Elevation gain westbound: 214 feet<br>Elevation gain eastbound: 1,342 feet<br>Difficulty: moderate - Shade factor: 35% |
| Bikes not allowed - Horses allowed - Dogs not allowed |
| East access: N34° 5.156', W118° 39.610'     West access: N34° 4.571', W118° 41.167' |

Backbone Trail. Many of the segments were pre-existing trails before they were combined to make the Backbone Trail. The old trails naturally had their own names and many trail users and hiking books still refer to them that way.

At the top of the chapter, you'll find the one-way **BBT segment length.** This mileage is the part of that chapter's hike that is *actually on the Backbone Trail.* It's most useful for through-hikers who are continuing directly from one segment to the next.

Next is the **day hike length**, which is sometimes longer than the BBT segment itself. This happens when the day hike requires additional walking on an access trail just to reach the Backbone Trail. If you are day-hiking just that one segment as described in the chapter, THIS is the actual mileage that you will hike. In any case, remember that if you intend to hike out-and-back, you must double your mileage.

We also list our choice for **suggested day hike.** Usually it's "east-to-west," but occasionally we describe the day hike going west-to-east, and sometimes we recommend doing it one-way only.

Next comes the **elevation gain**, listed both westbound and eastbound. For example, westbound elevation gain is the amount of feet you'll have to climb while heading that direction. It accounts only for uphill; downhills are not counted. If you return and head eastbound, you'll then have to climb the number of feet listed under the eastbound direction.

The **difficulty rating** of "easy," "easy to moderate," "moderate," "moderate to hard," or "hard" is determined by a combination of length, steepness, elevation gain and general tough going (such as boulder scrambling or a heavily rutted trail). An easy rating means the trail is suitable for the inexperienced hiker. A "hard" hike requires some stamina and shouldn't be attempted by a total newbie.

A **shade factor** follows. This statistic is the approximate percentage of the hike that is shaded. In Southern California, this number is important. On a warm, sunny day it can mean the difference between enjoyment and misery. The shade we're referring to is all-day shade from foliage cover. If you hike early or late in the day, expect more shade from steep hillsides, cliffs, overhangs, or foliage not directly overhead. Of course, wildfires can severely, albeit temporarily, lower the shade factor.

Regulations for **bikes, horses and dogs** are next. In general, dogs are allowed on leash in the National Recreation Area but are not allowed on trails in the state parks.

Bikes are allowed on parts of the Backbone Trail, but not all of it. They're legal on nearly all dirt roads. For the single-track sections, it's a little trickier. Most of the newly built Backbone sections in the National Recreation Area were co-created with the help of mountain biking organizations, so naturally those sections are open to bikes. For example, the entire 27-mile stretch between Las Virgenes Road and Triunfo Pass allows bikes. By contrast, the entire 17-mile single-track section from Eagle Junction (in Segment 2) to Las Virgenes Road is closed to bicycles.

Finally, we list **GPS coordinates of the trailhead(s)**, in case you prefer to get driving directions from your device instead of using our directions.

After the chapter header, the majority of each chapter is comprised of detailed trail descriptions. Within the description, you'll notice occasional footnote numbers. These are where junctions are mentioned, and they refer to the **GPS coordinates** of that particular junction. If you hike with a mobile device and you find yourself at one of the Backbone Trail's many confusing junctions, just check the list of GPS coordinates at the end of the chapter and you'll be able to verify which junction you are at. The trail description will tell you which way to go. Please note that GPS accuracy may vary by a few yards. Some devices' GPS functions don't work very well – make sure yours is trustworthy.

Occasionally, **side trips** are described in the chapter, written in italics within a grey box. These are mostly short detours to spots that we think are worth the extra effort. Be aware that some of these side trips may not be suitable for all hikers, particularly the thinly used overgrown paths or trips to places with cliff exposure.

You'll find another group of items also described in italics within grey boxes. These are **alternate routes** which bypass or parallel the Backbone Trail and rejoin it farther into the hike. An alternate route might be a more pleasant hiking option, such as a shadier or prettier trail, or it may just be a shortcut to return via an easier hike. The mileages described for alternate routes always start from the point at which you left the Backbone Trail.

Occasionally, we may also briefly mention an **alternate exit**; this is simply a quick "bailout" route to exit the trail to the nearest road access.

**Maps** and **elevation charts** are at the end of each chapter. Although our maps should suffice for most hikers, if you want more detail and topographic data, we recommend Tom Harrison Maps, which are waterproof and tear resistant. Their four maps covering the Backbone Trail ("Topanga State Park," "Zuma-Trancas Canyons," "Malibu Creek State Park" and "Pt Mugu State Park") can be found at most outdoor recreation shops and at tomharrisonmaps.com.

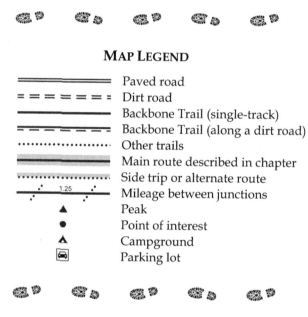

## MAP LEGEND

| | |
|---|---|
| ══════════ | Paved road |
| = = = = = = = | Dirt road |
| ▬▬▬▬▬▬▬ | Backbone Trail (single-track) |
| = − − − − − = | Backbone Trail (along a dirt road) |
| ···················· | Other trails |
| ▬▬▬▬▬▬▬ | Main route described in chapter |
| ···················· | Side trip or alternate route |
| ∴ 1.25 ∴ | Mileage between junctions |
| ▲ | Peak |
| ● | Point of interest |
| ⚲ | Campground |
| 🚗 | Parking lot |

## THROUGH-HIKING THE BACKBONE TRAIL

Although we generally recommend day-hiking the trail in segments, the Backbone Trail can be through-hiked and some hardy souls have accomplished it. This is by no means an easy feat. The current lack of adequate camping along the route presents the greatest logistical challenge. Water sources are scarce in some stretches but can be planned for. With the exception of Topanga Village, virtually no shops with food and supplies exist near the trail.

The one-way, east-to-west trek takes five to six days and can be accomplished in segments of no more than eighteen miles per day. There are also some "cheats" available (described later in this chapter) which can help to make the journey easier but may require additional days. For detailed description of the segments when through-hiking, consult this book's segment chapters, and in the case of segments which are described west-to-east, simply reverse the description details.

**Camping** is allowed only in designated campsites, and there are currently few campgrounds along the Backbone Trail. The National Park Service plans to create roughly a dozen more camps along the trail – some will be exclusively for equestrians, others solely for hikers, others for hikers and mountain bikers – but it will likely be a few years before the campgrounds are completed. When that occurs, the nature of through-hiking the trail will change dramatically. (And of course, we will update this hiking guide accordingly in subsequent editions.)

For now, most of the existing campgrounds are group camps which normally require ten or more people for a reservation. However, a group campground may grant you a waiver (allowing you to reserve a single site) if the camp is not already reserved and you're not planning to stay on the weekend. Contact the National Park Service or state park for reservations.

Campgrounds along or near the route are, from east to west:

- **Musch Trail Camp** at mile 10.2. A small 8-site campground right along the trail. No reservations; first come first taken.
- **Malibu Creek State Park Campground** at mile 24.6. A family campground with 62 tent or RV spaces. The campground is an additional 1½- mile walk from the Backbone Trail north on Las Virgenes Road.
- **Circle X Ranch Group Campground** at mile 51.1. A group campground with tent spaces, drinking water, pit toilets. The campground is an additional 1.7-mile walk from the Backbone Trail west on Yerba Buena Road.
- **Danielson Multi-Use Area** at mile 59.2. A group campground with water, flush toilets and showers located along the trail.
- Looking at this list, the weak link in it is clearly the stretch between Malibu Creek and Circle X, a distance of 26½ miles not including the walk to and from the campgrounds. To break this stretch up into two days, we need another source of lodging in the middle. One potential solution is the **Calamigos Guest Ranch** near the intersection of Mulholland Highway and Kanan Dume Road, 1.4 miles north of the Backbone Trail (at mile 36.1). The ranch is a multipurpose facility, hosting weddings, parties and other events. It also includes a high-end resort with cottages. It's not really the kind of place meant for backpackers – it's more aimed at the spa crowd – but they might accommodate you if you don't mind paying the lodging fee.

**The route**, if staying overnight at these locations, would be as follows: Starting at Will Rogers State Historic Park, hike west for 10.2 miles and spend the first night at Musch Campground along the trail. On day two, hike 14.4 miles to Las Virgenes Road, then walk another 1.5 miles north on the road to the Malibu Creek State Park Campground. On day three, return to the Backbone Trail, then follow it west for another 11.5 miles to Kanan Dume Road, then walk an additional 1.4 miles north on the road to Calamigos Ranch. On day four, return to the trail, then take it west for another 15.0 miles to the Yerba Buena Road crossing at Triunfo Pass, then walk another 1.7 miles west on the road to reach the Circle X Campground. On day five, return to the trail, then hike 8.1 miles to Danielson Multi-Use Area & Campground along the trail. On the final day, continue west on the Backbone Trail for its final stretch of 8.2 miles until it ends at the Pacific Ocean. (These final two days could be combined into a long day spanning 16.3 miles of trail.) The entire trek, including walking to and from campgrounds, would be 76.6 miles.

**Water** is difficult to obtain along certain parts of the trail. Piped drinking water is available at Will Rogers Park (mile 0), Musch Trail Camp (mile 10.2), Trippet Ranch (mile 11.1), Dead Horse Parking Lot (mile 12.2), Tapia County Park (just north of Las Virgenes Road Parking Lot at mile 24.6), Malibu Creek State Park Campground (mile 24.6 plus 1.7-mile road walk), Calamigos Ranch (assuming you stay there) (mile 36.1 plus 1.4-mile road walk), Circle X Campground (mile 51.1 plus 1.7-mile road walk), Danielson Multi-Use Area (mile 59.2) and the Backbone Trail terminus at the La Jolla Canyon Day-Use Area (mile 67.4) in Point Mugu State Park.

In addition, there are several stream crossings which may serve as sources of water, depending on the season. These are Topanga Canyon Stream (mile 12.3), Old Topanga Canyon Stream (mile 13.1), Dark Canyon (mile 22.4), Malibu Creek (mile 24.4), Upper Solstice Canyon Stream (mile 31.1), Zuma Creek (mile 37.1) and Trancas Canyon Stream (mile 40.5). Remember to always treat stream water before drinking.

Some hikers stash away cashes of water and/or food before hiking. Of course, you can't necessarily assume they'll still be there when you arrive.

**Restrooms** are few and far between. Toilets may be flush or pit, and they may not always be open. These can be found at Will Rogers Park (mile 0), The Hub (mile 7.6), Musch Trail Camp (mile 10.2), Trippet Ranch (mile 11.1), Dead Horse Parking Lot (mile 12.2), Las Virgenes Road Parking Lot (mile 24.6), Tapia Park & Malibu Creek Campground (north of Las Virgenes Road Parking Lot), Latigo Canyon Parking Lot (porta potty, mile 33.9), Kanan Dume Road Parking Lot (mile 36.1), Calamigos Ranch (assuming you stay there, north of Kanan Dume Road Parking Lot), Circle X Campground (mile 51.1 plus 1.7-mile road walk), Danielson Multi-Use Area (mile 59.2) and the Backbone Trail terminus at the La Jolla Canyon Day-Use Area (mile 67.4).

**Showers** are even fewer and farther between. You can find them at Malibu Creek State Park Campground, Calamigos Ranch and Danielson Multi-Use Area. You must be an overnight guest to use them, and the Danielson showers are some-times closed.

## ALTERNATIVE WAYS TO THROUGH-HIKE

If this all sounds like an overachieving week in backpacker purgatory, there are some creative alternatives which can allow for a less taxing through-hike, or at least a reasonable facsimile.

The first involves using **Airbnb** to find alternative lodging. Who says you have to camp, when Airbnb lists a surprising number of accommodations not far off the Backbone Trail? Cottages, apartments, semipermanent RVs, a retired old school bus, an airstream trailer, and tents for rent on private land are a few of the possibilities. Some of the accommodations include showers and kitchens, while others are no more than a place to crash.

If you'd prefer to let someone else handle the major grunt work, check out the Santa Monica Mountains Trails Council at smmtc.org. Every spring, they sponsor an **Annual Backbone Trek,** a seven-day group hike on which hikers carry only a daypack. The Council hauls the gear and provides van shuttles to and from camping areas. Their popular trek sells out months in advance, so be sure to plan far ahead of schedule.

*The Santa Monica Mountains Trails Council's Annual Backbone Trek*

Want a packaged, customizable through-hike? **Trail Magic Adventures** is a company that promises an "all-inclusive hiking and camping experience" along the Backbone Trail. For a fee, they provide a guide, meals, gear and privately-owned campsites. You provide the daypack.

There are also some ways to enjoy an "equivalent through-hike," offering the experience of hiking the entire trail without actually staying overnight on it. This can be nearly as fun as backpacking, with substantially less required in the way of roughing it.

One way to do an equivalent through-hike is to sign up for the Santa Monica Mountains National Recreation Area's series of **ranger-guided hikes** on the Backbone Trail, covering the entire trail over a period of eight Saturdays every spring. Check the National Recreation Area's website for more details.

Another group offering guided group hikes that cover the BBT is the Sierra Club's Wilderness Adventures Section. They offer an annual **Backbone Trail Festival** every fall, spanning the whole trail over four weekends.

Finally, if none of these alternatives appeal to you but you still long to enjoy something like a through-hike experience, consider the simple, yet effective **bedroom shuttle method.** An attendee at one of our book signings once described how she trekked the Backbone Trail over six continuous days while sleeping in her own bed every night. After she'd hiked a full day, she'd simply have a friend pick her up at a designated road crossing and take her home for the night, then drop her off the next morning at that same point to resume her "through-hike." Genius!

No matter how you approach it, a successful through-hike takes a great amount of preparation, planning and research – far more than can be provided solely in this book. For more detailed information, contact the National Park Service and California State Parks.

# A BRIEF GLOSSARY

Don't know what an "elfin forest" is? How about a "use path?" Here's a list of terms that we throw around in this book which might need a tad of explaining:

**BBT**: Abbreviation for Backbone Trail.

**Bushwhack:** A path so overgrown or faint as to be barely discernable, requiring you to push your way past twigs, branches or tall grass.

**Cairn:** A mound of small stones placed to mark a trail.

**Class I, II, III:** Levels of rock climbing difficulty. Class I is basic trail walking; II involves some scrambling over boulders with the use of hands; III requires a bit of actual climbing technique, the use of hands, and possibly a rope.

**Copse:** A small group of trees.

**Death march**: An unpleasant hike on a dirt road under oppressive sun on a hot day.

**Elfin forest**: Old-growth chaparral which is tall enough to resemble a miniature forest. It is taller than the average person and often creates a shady overhang.

**Flower sniffer:** A hiker who prioritizes the enjoyment of scenery and nature over getting to a destination; the opposite of a "tiger."

**Head (of a canyon):** The upper part of the beginning of a canyon. Frequently the Backbone Trail winds around the head of a canyon.

**Hoodoo:** A naturally occurring rock tower.

**Local high point:** Not the highest point on the entire hike, but the highest point for a substantial distance in either direction.

**Monkeypants:** This word does not appear in the book. We just threw it in to see if anybody actually reads these glossaries.

**Motorway:** Not to be confused with an English freeway, in California a motorway is a dirt road that is (or once was) suitable for motorized service vehicles.

**NPS**: Abbreviation for National Park Service.

**Out-and-back:** A hike on which you go to a certain turnaround point, then return the way you came. Its length is double that of a one-way hike.

**Peak bagger:** A hiker who focuses on conquering summits, often working from a list.

**Riparian:** Streamside; or the type of woods found along a creek or a canyon bottom.

**Seasonal:** Only running with water during the wet part of the year or after a rain.

**Single-track:** A hiking trail or path which is only wide enough for one pedestrian, bicycle or horse; not wide enough to be considered a dirt road.

**"T"-junction:** A junction of trails shaped like a T, where one trail ends perpendicular to the other trail.

**Termini:** Plural of terminus; it means both endpoints of the trail.

**Tiger:** A fast, aggressive or goal-oriented hiker; the opposite of a "flower sniffer."

**Use path**: A small, short and unmaintained path which is created simply by people tramping it down over time.

**"Y"-junction:** A junction of trails shaped like a Y, where one trail branches off into two diverging routes.

# CHAPTER 3:
# ON THE TRAIL

Most likely, a fun experience awaits you on any Backbone Trail trip. However, to paraphrase a common saying, "stuff happens." A little preparation and responsibility will greatly reduce the risk of "stuff" happening to you. The good news is that hiking properly requires hardly any more effort than screwing up royally. And while screwing up royally often leads to a good story, just remember that a good story is what you get when you had a terrible time. This chapter contains some tips, suggestions, warnings and advice to help keep your trouble to a minimum.

## GENERAL HIKING TIPS

The best time of the year to hike is almost always the spring, when plants are green, the falls have water in them, and the weather is comfortably cool. Late fall comes in at a close second, with weather in the same comfortable range, and with the fire season diminishing. Many days in winter are warm enough to be equally appealing, and even the coldest days are merely on the crisp side. Early summer mornings can be workable when the "June Gloom" keeps the sun off the hike until midday. The worst months to hike the Backbone Trail are July through mid-October, when fire season threatens, and the strong sun and oppressive heat can turn an otherwise pleasant hike into a "death march."

In general, the best way to beat the heat is to hike in the morning. This works particularly well when your hike heads uphill to a turnaround point (like a summit). As you expend your energy hiking uphill, the sun is low and the temperature cool. Then, as the sun strengthens and the air heats up, you are doing a much easier downhill walk. Beware of the inverted version of this hike, where you head down into a canyon in the cool hours and then are trudging uphill during the heat of the day. These types of hikes can work well near sundown – but be mindful of the oncoming dusk.

Consider whether the sun will be in your eyes. If you hike in the early morning, starting out in a westbound direction is better than eastbound. When the sun is at its lowest angle it will be at your back instead of right in your field of view.

When possible, hike on weekdays. Some segments of the Backbone Trail, particularly those near the metropolis, are fairly popular on weekends. While you will never feel like you're fighting off the crowds at Disneyland, parts of the trail can have enough users on a Saturday afternoon that you never really get that "alone" feeling you may be seeking. This is particularly true the closer you get to Los Angeles. We found that when we first hiked the trail – on weekends – there seemed to be a lot of mountain bikers. When we later re-hiked it – this time on weekdays – most of the bikes had disappeared.

Not only does a Saturday afternoon hike increase the chance of that "too many people" feeling, it also greatly increases your chance of what we call a "too many knuckleheads" vibe. While most users of the trail are perfectly responsible trailmates no matter what day of the week it is, there always seems to be that element of weekend warriors who don't know how to behave. Cellphone yackers, out-of-control dudes on bikes, yelling kids, and in-your-face teenagers seem to multiply on Saturdays and Sundays. Again, these folks tend to be more plentiful nearer the city, and while always relatively small in numbers, you are more likely to encounter them on the weekends.

Finally, come prepared. Don't just hit the trail without any necessities... or we'll lump you in with the aforementioned weekend warriors. You don't have to be Joe Gearhead to enjoy the Backbone Trail, and people do have a tendency to over-pack, but it makes sense to consider bringing at least some of the following items:

Hiking boots, water, trail map, trail description (i.e., this book), compass, Swiss army knife, cellphone, backup phone battery, hand wipes, first aid kit (or at least some bandages), flashlight (the rechargeable crank kind is the most reliable), convertible hiking pants and shirts, sunhat, sunblock, hiking socks, camera, lipbalm, baby powder or talcum powder, some food, and sunglasses. Carry it all in a fanny pack or knapsack (except the boots and clothes, of course!)

## HIKE HACKS

Feel like getting away from it all, but don't feel like busting your hump? If you're looking for a way to enjoy a Backbone Trail segment without putting in a six-, eight-, or ten-mile hike, there are plenty of "cheats" available to make a day hike shorter and easier.

First of all, consider doing half a segment. Nobody says you must hike the entire segment or you're some sort of flop. In many cases, the best features of a segment occur within the first quarter or third, making for a pleasant out-and-back hike to whichever turnaround point you choose.

However, if you really want to complete a segment the easy way, your best bet is to turn an out-and-back hike into a one-way hike of half the length. There are several ways to do this:

- **Use two cars.** One car (the pick-up car) is left at the endpoint. Then the hikers drive the other car (the drop-off car) to the starting point. When the hike is finished, the group pile in the pick-up car and drive to retrieve the drop-off car.
- **Use a car shuttle.** One volunteer in the group doesn't hike and instead drops off and picks up the hikers with the same car. (Of course, the volunteer can still hike in from the endpoint and meet the group halfway, joining them for the second half of the hike.)
- **Consider a rideshare service.** Instead of a volunteer shuttling you, pay Uber or Lyft to be the "shuttle." You'd be surprised how many segments we've hiked using a rideshare. If you can get cell service, they'll usually come.

IMPORTANT: Only use a rideshare BEFORE you start your hike (by parking your car at the endpoint and having the rideshare take you to the starting point). NEVER hike first and assume you can contact a rideshare when you are done, or you may be forced to turn around and hike back to your car if you can't get cell service.

- **Hike up & coast down.** On certain segments (3, 5, 10, 15 and 17), it is possible to leave a bike locked at the uphill end of the hike and coast most or all of the way back down to the starting point via paved roads. It's a judgement call; some folks love it while others find it unnerving. Exercise caution, as the mountain roads can be winding and steep, and not all drivers are looking out for bikes. Use a helmet, gloves, and good brakes, verify the road distance and elevation change on Google Earth or a topo map first, and – here comes the legal disclaimer – ride at your own risk.
- **Busses?** Hmm. Technically, you could use public bus lines as a shuttle on Segments 1 and 20. But we don't recommend it because they're unreliable and slow.
- A **mountain bike** trekked in and dropped off by a friend can be used to chop off the last three dirt road miles of Segments 19 and 20. A short bike ride along the shoulder of Pacific Coast Highway can also be used to finish the loop of Segment 20, necessitating only one car instead of two.

## TRAIL HAZARDS

Some hazards are serious, like sunstroke. Others are merely annoying, like being poked by the tip of a yucca. Some can be prepared for, like dehydration, and others come out of nowhere, like a fellow hiker's unleashed Rottweiler. Some you are probably already aware of, like sprained ankles, and others you probably hadn't thought of, like… drumroll… a coyote stealing your bag which contained your car keys! At any rate, you should be aware of the hazards out there in the great out-of-doors and prepare for them as best as possible. Here's a list, arranged in order of likelihood:

**Dehydration.** In California it is a must that you carry plenty of water. Note available drinking water sources before you hike. Drink in advance of thirst; by the time you feel thirsty you are already somewhat dehydrated.

**Sun exposure.** This could range from basic sunburn to heat exhaustion. ALWAYS bring a sun hat, which will keep you cooler and prevent a burned face. Wear sunblock, even on overcast days. In the event of heat exhaustion, rest in a cool, shaded place and drink plenty of liquid.

**Sprained ankles.** Wear hiking boots, ones high enough to re-enforce your ankle area.

**Foot blisters.** These are often caused by wetness in the foot or excessive friction. Wear hiking socks which wick away moisture, and if you're prone to blisters, carry an extra pair of dry socks. Consider using moleskin to protect blister-prone areas.

**Poison oak.** Learn to recognize its well-known pattern of three shiny leaves, which may be green or red. Take particular caution in shady streamside areas where it loves to grow. If touched, wash the area immediately if possible. Try to avoid touching the outside of clothes which may have rubbed up against it. When home, wash any exposed clothes in hot water and detergent.

*Poison Oak*

**Getting lost.** Although the Backbone Trail has plenty of opportunities to get on the wrong track, it's hard to become hopelessly lost (as in "Hansel and Gretel meets Blair Witch" lost) on the BBT. More likely, you'll make a wrong turn and end up having to backtrack. Use a hiking guide (like this one) and adequate maps. A compass and GPS device (both are probably on your phone) are useful as well. Although much of the Backbone Trail has limited or no cell reception, most phones will receive GPS data on it. Make sure your phone is charged in case you need to make an emergency call.

**Ticks.** These little fellas really suck. Literally! Contrary to popular belief, they don't usually drop down onto you from trees. Instead, they hop aboard from grass or shrub branches as you pass. Make sure to check for them after a hike.

**Bushwhacking.** This means having to shove your way through a heavily overgrown mess of twigs, tall grass or brambles. If you stick to the Backbone Trail you probably will never have to bushwhack, but some of the side routes or an unanticipated detour may require a bit of it. The most common hazard from bushwhacking is a bunch of scratches. A bushwhack is a likely place to end up with ticks or a sprained ankle as well.

**Rattlesnakes.** They are plentiful in the Santa Monica Mountains. Although we've seen many of them, we have never been bitten. It's pretty easy to avoid being bitten by a snake. They're often found splayed out across the trail, almost trying to be in the way. Snakes tend to hang out in sunny, rocky areas where they can soak up the rays to regulate body temperature. If you see one, back off a good twenty feet or so. Gently toss some pebbles at it until it slithers off the trail, and only then pass with caution.

**Wet feet.** Getting your feet wet isn't the end of the world but hiking a long distance with a soaked foot can lead to blisters. There are plenty of stream crossings along the Backbone Trail, and in rainy periods some of them can get a bit dicey. If possible, change into dry socks or just chill while your socks and boots dry out in the sun.

**Nightfall.** This is a "hazard" many folks don't think about. People always assume they'll be back at their car well before dusk... until something goes wrong. They find themselves far from the trailhead with no flashlight and no moon as darkness creeps into the forest. People begin to get panicky, people hurry too fast, people trip and sprain an ankle, and then other people have to help them back to the car and wind up promising to do the dishes for a month because they feel bad that they didn't bring a flashlight, even though the other people could've thought to bring one... Of course, we have no idea who these "people" are.

**Rapid weather change.** In the Santa Monica Mountains, you don't really have the threat of a dangerous weather shift like you might on a place like Mount Baldy. The worst you'll likely endure is wind, rain or fog coming in unexpectedly. On a chilly day with wind-blown fog, however, these mountains can get surprisingly cold, so be sure to bring zipper-convertible pants and shirts which can cover your arms and legs. In the winter, bringing a light jacket isn't a bad idea.

**Hiking alone.** This in and of itself poses a potential hazard. It's advised to always hike with a partner, and if you must go it alone,

*Bad weather coming in fast over Sandstone Peak*

leave behind details as to where you went. Otherwise, you might end up like that guy who had to cut his arm off in the film *127 Hours*, which is the ultimate example proving our adage, "a good story is what you get when you had a terrible time."

**Mountain bikes.** We're sorry to have to include this as a "hazard," as the vast majority of mountain bikers are respectful trail allies. But there have been incidents of reckless mountain bikers (usually the weekend warrior types or teenagers imitating extreme sports) who lose control and injure others with the bike or with flying gravel. It's best to keep an ear out for them, as these types usually give away their position via rattling equipment, spitting gravel, or just plain yelling.

**Falls.** By this we mean something worse than tripping and spraining an ankle. This is more of the "off a cliff" type of fall. A little caution goes a long way in avoiding these. The Backbone Trail stays generally clear of major precipices, but there are some out there, particularly on side trips. If you are the unsure-footed type, avoid these areas as described in the book. Be particularly careful around ledges and crumbly slopes. Ice on the trail can pose an uncommon slipping hazard; a much more frequent cause is slippery gravel. Don't let yourself fall into the trap of hiking while looking down at something like your phone or a map.

**Fire.** Wildfires can happen anytime, but they're more likely during the summer and fall. Always heed posted fire closures, and never do anything that could start a fire. If you're particularly concerned about fires, you might want to avoid hiking on warm, windy summer and autumn days.

**Other dangerous plants.** Although poison oak is the superstar of the evil plant world, we do have some other nasty plants who would just love to poke you with pointy tips. Cactus, succulents like yucca, brambles and even thick twiggy chaparral can cause cuts – especially if you fall into them. A small first aid kit comes in handy in situations like these.

**Car break-ins.** This is a trail hazard that gets overlooked, mainly because it doesn't actually happen on the trail. Hikers sometimes leave their cars vulnerable to theft at trailheads because they think they have gotten away from all the city crime and they're feeling safe. Unfortunately, certain thieves prey on hikers' cars. We personally witnessed a sad scene at the Triunfo Pass Parking Lot when we returned from our hike to find a crying couple and a cop next to their smashed car windshield. For some reason, the man had left his wallet sitting in the car. Perhaps he didn't want any predatory animals stealing his credit cards along the trail.

**Large wild animals.** Mountain lions! Coyotes! Bobcats! Killer deer! Although some visitors find wild animals worrisome, in general, they'll want nothing to do with you. Most likely, if you see one of these animals it will be as it hurries away from you; in this case you should consider yourself lucky to have seen it at all. Franky, we wish there were more of them left in the Santa Monica Mountains. In the infinitesimal chance that you do encounter a mountain lion acting aggressive, make yourself appear larger by raising your hands, make plenty of

*Does a mountain lion qualify as a hazard? If it's THIS close, it does!*

noise and slowly back away. Never turn your back or run. If attacked, fight back, protecting your neck and throat.

**Flash floods.** Pretty uncommon in the Santa Monica Mountains, but they do occasionally happen. If you are hiking along a ravine or canyon bottom and a flood starts forming, immediately climb to higher ground.

**Rock climbing.** It's only a hazard if you weren't planning on doing it. The Backbone Trail requires no rock climbing, and a few of the side trips or alternate routes require a small amount of class II scrambling up boulders on all fours. If you see an enticing cliff along the trail but are not prepared to do rock climbing... don't do rock climbing.

# TRAIL ETIQUETTE

Your goal, we would hope, is to enjoy the outdoors without spoiling it for someone else. Being a responsible trail user is mostly a matter of common sense. However, there are some rules which don't always occur to everyone intuitively, so we'll give them a quick mention.

**Don't cut switchbacks.** To some this may seem like a victimless crime, but it leads to erosion and destruction of the very trail which you are enjoying.

**Don't litter.** Alright, we know you don't litter – you're not a complete jerk. But is tossing an apple core littering? What about tossing a bit of a Snickers bar? These are organic, right? Actually, they are litter, because they're considered non-native. Although the core and candy bar will decay, the former contains seeds of a non-native plant. We already have enough invasive plant species messing up the mountains, so don't add to it. The second – the candy bar – will never grow into an invasive Snickers bush (as delicious as that sounds), but a wild animal can and probably will eat it, causing it to fill up with junk food that provides no nutrients and may compromise its immune system. Basically, follow the rule "pack it in, pack it out."

**Don't take specimens.** Every pinecone or cool-looking rock that you take is one less pinecone or cool-looking rock for other hikers to enjoy spotting.

**Keep the noise down.** Other people came here to get away from it all. Don't bring "it all" back to them by spoiling their enjoyment of some peace and quiet by yacking loudly with a hiking partner or, even worse, into a cellphone.

**Don't smoke.** Jeopardizing your health is your own private business, but accidentally starting a wildfire is not. If you smoke along the Backbone Trail, you're jeopardizing miles of scenic beauty and countless homes for the sake of a cig.

**Be responsible with the kiddies.** Some folks think it'll be great to take the little ones on a walk in nature… but they have no clue what to do with little ones in nature. They let their kids run amuck, breaking branches, throwing rocks, screaming. We remember one particular lady whose kids were running down the trail, kicking up dust and yelling like crazy. When we pointed out that her kids were disturbing the peace and quiet, she told us, "Oh that's good. It'll scare the snakes away." We didn't have the heart to tell her that snakes have no ears.

**Use those brakes.** Don't be one of the 10% or so of mountain bikers who threaten to give the rest a bad name. Remember that the Backbone Trail is a public, multiuse trail, and not anyone's personal, private thrill ride. For instance, shouting "on your left!" while going double the trail speed limit doesn't somehow make it safe.

**Keep Sir Wagsalot on a leash.** As much as most dogs love a good hike, they are not allowed on much of the Backbone Trail. Where they are allowed, they must always be leashed.

**Dim it down.** If hiking after dark, please resist the urge to use ultrabright LED headlamps and tactical flashlights. They may be great for working a coalmine, but in the mountains after dark, they disturb wildlife and degrade the night experience for others. Use low-lumens headlamps or flashlights instead.

# Chapter 4:
# Flora, Fauna & Faultlines

Many hikers, in search of scenery and recreation, take to the trail but never consider what it is that they are hiking through. Your authors are no exception; it took us years of enjoying the outdoors before we started actually wondering, "What is all this stuff we're passing?" For example, it was longer than we'd care to admit before we discovered that chaparral was not the name of a plant, and that any area of chaparral wasn't simply the same plant species repeating over and over.

The more you learn about nature, the more enjoyable the hike becomes. For the intensely curious, there are volumes out there filled with in-depth information about the geology, flora and fauna of the Santa Monica Mountains. If you were so inclined, you could spend weeks poring over the minute details of the science behind these mountains. Our chaparral plant community alone has entire books devoted to it, and one can purchase a copy of a detailed report describing the mineralogical makeup of just the *eastern half* of the Santa Monica Mountains. But unless you're a geologist or a botanist (and maybe even if you are), you'd probably rather spend more time out on the trail than you would reviewing statistics.

For the average hiker, a quick description will suffice. With that in mind, here's a very brief introduction to the area to help enhance your enjoyment of the trail.

## Geography

The Santa Monica Mountain Range covers a long, generally rectangular area west of Los Angeles, bounded on the south by the Pacific Ocean, on the north by the San Fernando and Conejo Valleys, on the west by the Oxnard Plain and on the east by the Los Angeles River.

One quarter of the range bisects the city of Los Angeles and is often referred to as the Hollywood Hills. This section east of the 405 freeway is lower and less rugged than the rest of the range, and since it has been heavily developed for some time, the Backbone Trail does not run through here. Griffith Park and the Hollywood Sign are at its far eastern end.

The rest of the range is wilder, higher, and more rugged. Pacific Coast Highway runs along its southern border through Malibu; the 101 freeway marks its northern border through the San Fernando and Conejo Valleys. The BBT generally follows the crest of the range, paralleling the two major highways roughly halfway in-between.

Several major canyons cut into the range's southern slope to the ocean, many with roads following them. These major canyons are, from east to west: Cahuenga, Laurel, Franklin, Sepulveda (containing the 405 freeway), Mandeville, Topanga, Malibu, Solstice, Zuma, Trancas, Arroyo Sequit and Big Sycamore. The canyons draining northward are much shorter than the south-draining canyons. Of all the major canyons, only Malibu Canyon cuts all the way through the range and drains the region north of it, bisecting it rather neatly into two halves near its midpoint.

The peaks gain altitude as you head west, culminating in 3,111-foot Sandstone Peak near the western end. They also get more rugged and craggy as you go west.

The Santa Monica Mountains are a part of the Coast Ranges, an aptly named series of mountains which run north-south near America's west coast. "But wait," we hear you say on cue, "I thought the Santa Monicas went east-west, not north-south." That's because the Santa Monicas are part of a smaller section (or *cordillera*) of the Coast Ranges known as the Transverse Ranges, meaning that this particular section runs perpendicular to the rest of California's mountains. A kink in the San Andreas Fault makes it turn temporarily east-west, and the mountains turn with it.

This non-conformist group starts with the Santa Ynez Mountains near Santa Barbara, then continues eastward with the San Rafael, Sierra Madre, Topa Topa and Santa Suzanna Mountains, then the Santa Monica Mountains, and finally the San Gabriel and San Bernardino Mountains northeast of Los Angeles.

The cordillera's east-west orientation is extremely rare for the Americas. In fact, the Transverse Ranges are considered to be one of only two east-west mountain ranges in North America. (The other one is the Uintas in Utah.)

## GEOLOGY

Geologically speaking, the Santa Monica Mountains are relatively young – pushed up about 10 million years ago – although the rocks which comprise them are far older and were laid down by ancient sediments long before they were elevated. The Malibu Coast Fault, running under Malibu itself, caused the uplift as the Pacific Plate scraped northwards against the North American Plate.

At one time these mountains rose to 10,000 feet, rivaling the height of today's Rockies. But since the majority of the range is comprised of crumbly sandstone, it eroded easily under the stress of rains, plant growth, temperature extremes and even wind. This erosive quality explains the range's rounded peaks, its numerous canyons, and the sandy and gravelly character of much of the Backbone Trail.

The sedimentary rock underlying the mountains comes in many layers created during different geologic periods, each having its own features. Some hold marine fossils. Others contain a conglomerate of pebbles. Others have a distinctive red, orange, purple or pale green coloring. You'll

*Backbone Trail below Sespe Formation sandstone pinnacles*

cross these different layers as you explore the Backbone Trail. The age of a layer has little bearing on whether or not it appears on the surface, since earthquakes and erosion have exposed different layers in different places.

In the western portion of the mountain range, you'll come across dramatic areas of buttes, rock towers and drop-offs. It looks different than the rest of the range and there's a good reason. These areas are not sandstone. They are igneous (volcanic) rock which intruded into the sandstone about 15 million years ago. Once the molten rock shoved its way in, it hardened with the sandstone still around it. Then, after the area was uplifted millions of years later, the wear-and-tear of erosion wiped away the soft sandstone, leaving the harder volcanic rock looking like it had been placed there after the fact.

A similar process happens on a smaller scale with hoodoos – rock towers – seen from time to time along the trail. These smaller, pointy curiosities are not volcanic. They are usually sandstone, just like everything around them, but they are capped with a somewhat harder layer on top and are left to stand taller than the surrounding landscape as wind and water slowly erode their bases.

Pinnacles can also resemble rock towers. They're not hoodoos, however, because their vertical orientation is due not to erosion, but to earthquakes. As the rock layers were pushed up by geologic force, they cracked, buckled, and were tilted upwards, leaving the once flat layers jutting into the air.

The easily eroded sandstone is also responsible for the occasional flat valley that you'll cross. These areas, such as the bottoms of Trancas, Big Sycamore and Rustic Canyons, are filled in with sediment washed down from higher ground.

Unless you're a major rock nerd, most of the geologic layers won't be of much interest to you as you hike. They tend to look pretty similar to the untrained eye. A few exceptions, however, are easy to spot and stand out:

The oldest and lowest layer is the **Santa Monica Slate**, deposited about 150 million years ago. It surfaces mainly under the Hollywood Hills region of the Santa Monica Mountains, so not much is exposed along the Backbone Trail.

The **Coal Canyon Formation**, a layer of sandstone, siltstone, and pebble conglomerate, is about 60 million years old. It's visible in the exposed walls of Topanga Canyon. Laid down by marine sediments, it often contains ancient mollusks.

**The Sespe Formation**, about 30 million years old, is the celebrity of our local rock strata (if such a thing exists)

*A volcanic intrusion on Boney Mountain*

and is by far the most colorful and easiest to spot of the layers. Comprised of sandstone and mudstone deposited in non-marine floodplains during a time when the ocean retreated, it often stands out as pink, purple, orange or red. You can thank Sespe for many of the hoodoos, cliffs, and colorful boulders along the trail.

Immediately above the Sespe Formation, the **Vaqueros Formation** comprises the next newest layer, dating back about 20 million years. This layer sometimes contains ancient shellfish fossils like Turritella, great examples of which can be seen along the Mesa Peak Motorway section of the Backbone Trail. The **San Nicholas Member** of the Vaqueros Formation surfaces occasionally as pale blue or light grey rock, making for some picturesque cliff faces and ledges.

More recent are the **Topanga Formations** (created about 15 million years ago), a sort-of layer cake trio of sedimentary, then volcanic, then sedimentary rock. The huge white boulders strewn on the north slope of Saddle Peak are from this group.

Included roughly in this same time period are the **Conejo Volcanics:** that large infusion of igneous rock which we mentioned earlier, mostly prominent in the western half of the range. These rocks, along with the Sespe Formation, are the easiest to spot as they stand left behind after the surrounding sandstone eroded away. Up on Boney Mountain these rock protrusions stand all around like sentinels watching over the area. They are mostly andesite and basalt and are often mixed in with the sandstone of the Topanga Formations.

## PLANTS

If you've never viewed the Santa Monica Mountains from any closer than the freeway, you might dismiss them as a boring range covered by essentially the same thirsty-looking bush a million times over. But the closer you explore, the more you'll realize that nothing is further from the truth. These mountains offer an amazing diversity of vegetation. Several completely different types of plant communities thrive here, from deep forests to cactus-dotted canyons to windswept grassy summits to jungle-like patches of reeds. Even the ubiquitous chaparral has a surprising number of different plants calling it home, instead of the "same thirsty bush a million times over" that it initially appears to be.

So, what are plant communities? Think of each as a neighborhood of plants that all prefer whatever their particular locale has to offer: specific combinations of water, sunshine, temperature, and soil nutrients. The plant communities that you'll encounter along the Backbone Trail are (in order of abundance): chaparral, oak woodland, coastal sage scrub, riparian woodland, California grassland, oak/sycamore savannah, freshwater marsh, and non-native planted areas.

To understand the different plant communities, you must first consider the local climate that governs what grows here and where it prefers to live.

Our climate is known as a Mediterranean climate. Less than three percent of the world's landmass shares this type of climate, categorized by dry summers, cool and rainy winters, and relatively mild temperatures. This climate exists only in the Mediterranean, southwest Chile, southwest Australia, southern Africa, and here in

SoCal. The plants preferring it are ones which can go for long periods without water, are adaptive to fire, and don't have to endure freezes.

Within the larger climate zone are microclimates, which may be hotter or cooler, wetter or more arid, sunnier or shadier. For instance, the inland mountain slopes get less rain and fog than the coastal zone, leaving them prone to more extremes and less moisture. Southern and western-facing slopes receive stronger sun. Canyon bottoms collect more water. Summits endure harsher winds. Each of these areas hosts a different plant community that suits it best. The more you explore the Backbone Trail, the more you will discover these different communities. Let's take a closer look.

*Chamise (including bloom)*

**Chaparral:** Everyone knows about chaparral. It's that dry, bushy stuff, right? But what's it comprised of? Most hikers and bicyclists spend hours passing it by but have little clue as to what it is, if they are even aware of it at all.

Chaparral isn't the name of a single plant... well, not exactly. The name comes from *chaparro*, which is Spanish for scrub oak, one of the more than fifty plants that we collectively deem chaparral. But chaparral needn't be complicated; most of what you need to know about it can be summed up in a James Taylor title: *Fire and Rain*.

Rain (or the lack thereof) is what puts chaparral where it is. This wannabe forest thrives in an area too dry for a real forest and too wet for desert cactus. The drought-adapted chaparral plants are experts at conserving moisture through non-deciduous leaves that are either spindly (like pine needles) or waxy (like succulents), along with advanced root systems for deep drinking.

The other major factor is fire. Chaparral burns frequently and is used to it. The average area of chaparral burns every ten to thirty years. It's just the nature of these semiarid, sun-facing slopes. Fast growing chaparral has evolved to not only respond to fire, but to actually take advantage of it. Some plants sprout new growth from the burned stump, while others drop seeds that only germinate after fire.

Some chaparral manages to avoid fire for such a long period that it grows tall enough to resemble a forest. We call these areas *elfin forests*, where the plants can get up to fifteen feet tall, enough to shade hikers and create the occasional tree tunnel.

Whether mature or fledgling, chaparral tends to grow thick and impassible. If you try to bushwhack through it, you'll end up with more scratches than an alley cat in a fight with a raccoon. Cowboys quickly learned this and created leather leg armors to protect themselves, naming them "chaps" after the scratchy chaparral.

In order to find chaparral interesting instead of a snoozefest, you needn't be familiar with all of it. Most chaparral is comprised of a handful of old standbys:

# CHAPARRAL: AN INTRODUCTORY CHEAT SHEET

**CHAMISE** is the most common chaparral inhabitant. It adores the sunny environment and in a few instances you may spy a hillside that seems to be covered with nothing but chamise. It's easily identified by its short, spindly, needle-like leaves. At first glance it almost resembles a sort of puny pine tree or a thinned-out rosemary bush.

In rarer instances, you'll find chamise's more picturesque cousin, **RED SHANK,** mostly in groves along the western part of the trail. It too has needle-like foliage, but the needles are longer, softer and wispier. Its heavily peeling red bark is a giveaway.

The next most common chaparral plant is **CEANOTHUS**. More tree-like than chamise, the ceanothus' leaves are oval, paddle-shaped, waxy, and its trunks often grow twisted around each other in a latticed, rope-like pattern. Heavy thickets of ceanothus are the predominant component of tree tunnels over the trail. In spring, it blooms with a profusion of white or blue ball-shaped blossoms, making for an unforgettable sight.

**SCRUB OAK,** a scraggly little oak tree (or oak shrub) which rarely gets taller than fifteen feet, can look a bit similar to ceanothus at first glance. It can be distinguished by its leaves, which have small points on them instead of the ceanothus' smooth oval leaves. Look for little acorn caps on the branches – a telltale giveaway of scrub oak.

**LAUREL SUMAC** and **TOYON** are the next two most commonly found shrubs. Although not related, the novice might confuse them as they both are similar in height (five to fifteen feet) and have leaves that are long and canoe-shaped with pointy tips. The laurel sumac's leaves, however, have a distinctive red vein down the middle and grow in a cupped shape, while the toyon's are flat with mildly serrated edges. Toyon sports red berries in the winter, making it the closest thing SoCal has to a holly bush. Toyons growing wild in the hills above Los Angeles are where Hollywood got its name.

**MANZANITA** is one of the easiest chaparral shrubs to spot. Its beautiful red or brown bark is smooth and layered, looking like someone dipped it in melted chocolate. Its name means "little apple" because the manzanita produces miniature apple-like fruit, which can be eaten.

Of the smaller, non-shrub plants, **SAGE** and **MONKEY FLOWER** are two of the most common. You'll find both along the trail down near your knees. They may look similar to the untrained eye – long thin flat leaves about an inch or so in length – but one touch of the underside of the leaf will return a positive ID. Monkey flower feels sticky to the hand, while sage will make your fingers smell like turkey dressing. In the spring, the profuse orange blooms of the monkey flower are a lovely sight.

**SAGEBRUSH** is, oddly enough, not the same as sage. This waist-high, drought-tolerant plant often grows amongst sage, but its leaves are spindly and needle-like. Its pale green or smoky gray color makes it stand out from the darker green of its neighbors.

**DODDER** is sometimes found along the hotter, drier stretches of the trail. It's an easy plant to spot because (to us at least) it always looks like something that died. It has no leaves, appearing like a tangled mess of orange, yellow or brown spaghetti. This goofy-looking parasitic plant sponges off the life of other plants beneath it.

**YUCCA** is familiar to many Southern Californians, and it grows happily in the Santa Monica Mountains. Its tall single stems sporting white flowers stand above everything else on the arid hillsides, resembling glowing candles from a distance.

After chaparral, the next most common plant community found along the trail is an **oak woodland**. These beautiful and shady forests, comprised mostly of California coast live oaks, grow in areas receiving less sun, like north-facing hillsides. Coast live oaks are drought-tolerant and, unlike most other trees in the region, needn't live right by a stream, so their woods often sprawl up canyon slopes far above the streambeds. The woods can be thick with shade or only partially canopied, allowing plenty of other plants to grow beneath them. Squirrel, deer and birds love the acorns. Oak woodlands can also be found covering flat canyon bottoms if the area is wet enough. The old oak woods covering the bottom of Trancas Canyon are a great example.

*Manzanita's bark looks like it's been dipped in chocolate*

Deep and gloomy bay tree woods (not its own plant community but part of an oak woodland) often cover old land-slides and frequently comprise the shadiest part of a hike. You can tell a bay tree by rubbing one of its leaves and giving your fingers a sniff. The distinctive smell of bay leaf seasoning used in cooking is unmistakable.

**Coastal sage scrub** is another frequently encountered plant community. Often intermixed with chaparral or growing adjacent to it, coastal sage scrub takes over when things get a little too hot and dry for chaparral. This plant community is not like the desert with its harsh extremes found farther inland – it instead prefers the semiarid coastal climate found on south-facing slopes below 2,000 feet. As the name suggests, this community features plants which are lower, smaller and generally more scrubby looking than those of chaparral. Dry grasses, sagebrush, sage, California buckwheat, some cacti and succulents, as well as some smaller toyon and laurel sumac prefer life here.

*Toyon on the left; laurel sumac on the right*

**Riparian woodlands** are perhaps the most welcome delights on the trail. These brief sections of intense greenery along streams and canyon bottoms host big, beautiful trees with plenty of shade and a damp, lush feeling, in stark contrast to the dry reputation of Southern California. Sycamores, alders, oaks, some willows and cottonwoods all can be found here fighting for the water. Under the canopy grows an abundance of shade-loving plants like ferns, moss, clover, poison oak, mushrooms, even wild grape vines. Diverse wildlife enjoys feeding and living in the area – frogs can be particularly audible in the late afternoon and evenings.

**California native grasslands** are rare but wonderful. These open meadows are evocative of rural prairie lands. The "native" element is mostly gone, plowed under for farm or ranchland long ago, and whatever was left has been heavily decimated by invasive species such as mustard and thistle. Still, walking through one of these meadows makes for a special delight. The La Jolla Valley Natural

*It ain't dead, it's dodder!*

Preserve and Musch Meadows are two large, prominent examples along the Backbone Trail. Many other smaller fields may surprise you as well.

An **oak or sycamore savannah** can be found in a few places along the trail. It's a wide and flat valley plain which hosts mature trees, not in a compact forest setting but spread out amongst a sprawling grassland or scrubland. The area is only dotted with the occasional stately and full-grown tree, giving it a park-like feeling. The large sycamore savannah along the bottom of Big Sycamore Canyon (Segments 19 and 20) is the most impressive example, seeming to go on forever. A smaller oak and sycamore savannah exists along the bottom of Rustic Canyon as well (see Segment 1).

**Freshwater marshes** on the Backbone Trail can only be found in tiny patches wherever water collects in streambeds. For a few fleeting seconds you may pass by a thicket of plants that seem very out of place in these mountains, including pond lilies, cattails, tall grass and reeds.

**Pine forests** don't exist naturally in the Santa Monica Mountains, but you'll occasionally come across a copse of conifers which manage to hold on in the arid environment. When you do, it will actually be as part of our final plant community:

**Non-native planted vegetation** can be a surprise. Small groves of eucalyptus, pepper, palm or pine trees are picturesque, but they are not there naturally. They're left over from the old ranching days or were planted to accompany cabins or homesteads which are long gone.

# ANIMALS

Although you won't see most of them, there are over 450 animal species in the Santa Monica Mountains. The ones you don't see are, of course, the ones you want to see the most; lizards, squirrels and crows don't really cut it in the "guess what I saw today" category. Most of the critters along the Backbone Trail make it their business to stay unnoticed, but you're more likely to see their tracks or their scat. If you're lucky – and quiet – you will eventually catch a glimpse of some of them as they saunter or scurry away.

The larger predators like **mountain lions, bobcats** and **coyotes** remain the most elusive. With a few dozen mountain lions in the Santa Monica Mountains, you're lucky indeed to spot one... just not too close! One of your authors, Doug, was fortunate enough to spy a mountain lion for about half a minute up on dirt Trancas Canyon Road, far from any development. The cougar took a long glance at him, then languidly walked away and disappeared behind a curve. When Doug cautiously rounded the curve a few minutes later, it had vanished into the brush.

You're more likely to spot a coyote than the elusive wild cats. The "yip-yip-yowl" of the coyote at dusk can be a little disconcerting at first.

And as for **bears**... they don't live in the Santa Monica Mountains. According to the Fish and Game Department, any bears spotted in the region have wandered in from other areas. At any rate, black bears are

*Coyote*

non-confrontational and generally want to avoid you. The larger, more aggressive California grizzly – the bear seen on our state flag – was hunted to extinction about a hundred years ago.

The biggest animal that you are most likely to spot is the **California mule deer**, as they number several thousand in Los Angeles County alone. Like the wild cats and coyotes, they usually play keep-away, but a few herds in Trippet Ranch (part of Topanga State Park) and Will Rogers State Historic Park are accustomed enough to people that they'll hang around as long as you don't get too close or noisy.

Smaller creatures, often prey for the larger predators, can be more easily found: **rabbits, raccoons, squirrels, chipmunks** and the gerbil-esque **California kangaroo rat** are the most common rodents. **Grey foxes** can also be found in these mountains. Birds are plentiful and varied; some raptors like the **golden eagle, owl,** and **prairie falcon** either live in or visit the Santa Monica Mountains.

The **Pacific tree frog** is commonly found near streams, and while hard to see, its wonderfully rural croaking sound at dusk is hard to miss.

**Lizards** include the Western Side-blotched, Western Fence, Southern Alligator, Blainville's Horned and California Whiptail varieties. These little guys are all harmless and nonpoisonous, which is good because they seem to be every-

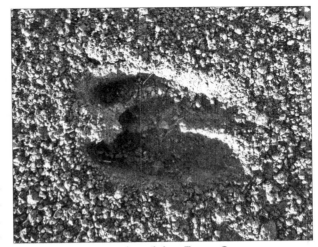

*Deer track in the mud on the trail above Trancas Canyon*

where along the trail, darting away as you pass through sunny, rocky areas.

These same areas are prime spots for snakes. Probably the most beautiful snake you'll see (and we rarely throw around the term "beautiful snake") is the non-venomous **California mountain kingsnake**, with its spectacular tricolored bands of red, white, and black that virtually scream "notice me!" However, the most notorious snake (and the one you probably pictured when you read the word "snakes") is the **southern Pacific rattlesnake**. Not to be confused with the similar-looking but non-venomous **gopher snake,** it can be identified by the diamond-shaped dark patches on its skin, and of course by its rattle if threatened. They're not particularly aggressive, however, and most will slither away when they sense your presence. If not, try gently tossing a few pebbles at it from a safe distance.

We once had the opportunity to witness the eerie ritual of two rattlers mating on the Murphy Stairs, a dilapidated pair of unbelievably long outdoor staircases leading down into Rustic Canyon (see Segment 1). The sight and sound of the snakes writhing in the air, half off the ground, rattles shaking furiously, occasionally tumbling down the staircase, was enough to send us home with a serious case of the heebie-jeebies.

*Lizards seem to be everywhere along the Backbone Trail*

# WILL ROGERS PARK TO THE HUB
### ("ROGERS TRAIL")

**BBT segment length: 7.6 miles**
**Day hike length: 10.2 miles one-way**
**Suggested day hike: west to east, one-way**

**Elevation gain westbound: 2,184 feet**
**Elevation gain eastbound: 858 feet**
**Difficulty: hard – Shade factor: 15%**

**Bikes allowed (except on Betty Rogers Trail section)**
**Horses allowed – Dogs not allowed in state park**

East access: N34° 3.259', W118° 30.803'     West access: N34° 7.880', W118° 33.190'

This lengthy segment, crossing the spine of the Santa Monica Mountains from the San Fernando Valley to the sea, spends much of its time following old bridle trails designed and ridden by Will Rogers back when the area was his property. Will, the humorist, radio personality and actor from Oklahoma who famously claimed, "I never met a man I didn't like," enjoyed mega-star popularity in the twenties and early thirties. At the trail's end in Will Rogers State Historic Park, you can still visit the home of the man who was renowned for home-spun quips like "A fool and his money are soon elected," and "Everybody is ignorant, only on different subjects."

This is one of the longest uninterrupted stretches of the Backbone Trail, making it a real hump-buster with a 1,600-foot descent towards the ocean. As such, we recommend taking this hike one-way, heading west to east, so that you are mostly descending and taking in increasingly spectacular views as you proceed. To do so, you'll need two cars (one parked at each end of the hike), a friend to shuttle one car from the starting point to the end, or a rather pricey rideshare trip (see Chapter 3).

The first half of this hike is somewhat ho-hum by Backbone Trail standards: sunny, dry, and at times repetitive. The upside to the first half is that you'll probably feel like you've really gotten away from it all, even though you are right on the edge of the second largest metropolis in the country. Once when we hiked this segment on a warm spring Sunday, we came across barely a single soul on the first half of the hike. Perhaps that's because it's gravelly, a bit overgrown in sections, and plenty of it is on sun-exposed fire roads.

Ah, but the second half! The second half more than makes up for the first. As you descend towards Will Rogers State Historic Park on aching feet, you'll be rewarded with unforgettable panoramas of miles of gorgeous Pacific coastline, stretching from Santa Monica to the Palos Verdes Peninsula, as well as most of the city of the angels laid out before you. Catalina Island, the towers of Century City, endless freeways, distant mountain ranges and the omnipresent beautiful blue Pacific... it's the kind of hiking experience you can only get in Southern California.

You'll also cross the mildly hair-raising Chicken Ridge Bridge, enjoy great views of Rustic Canyon from the serene, aptly named Lone Oak, climb the highest

*Los Angeles seen from Rogers Ridge*

peak in this part of the Santa Monica Mountains, and quite possibly spot deer in Will Rogers Park. You can even visit Will Rogers' home.

Many hikers do not complete this entire segment, choosing instead to take a shorter trip starting and finishing at Will Rogers Park. They may do an out-and-back hike on the Backbone Trail, going as far as Chicken Ridge Bridge or Lone Oak before turning around (see "A Shorter Alternative"). Or they might make it a loop hike by ascending via the Rustic Canyon alternate route and then returning via the Backbone Trail along the ridge (see "A Loop Alternative").

**A special note:** Upon first glance at a Backbone Trail map, you might be tempted to hike the entire first continuous stretch of the trail – between Trippet Ranch and Will Rogers Park – directly on one very long day. You *could* do this, but we have instead divided that long trek into two easier segments by short-cutting in on a spur trail from Reseda Boulevard to The Hub, a lonely junction of fire roads that intersect the Backbone Trail in the middle of Topanga State Park. This shortens the trek to Will Rogers Park by a half a mile and, more importantly, shaves off nearly half of the elevation gain, making the eastern part of the full segment more manageable.

**Directions to trailheads:** First, park your pick-up car at the end of the hike in Will Rogers Park. (This location is also where you'd park if you use a rideshare service to shuttle you to the start of the hike.) To get there, take Sunset Boulevard westbound from the 405 freeway and continue for 4.6 miles. Turn right onto Will Rogers Road and wind your way up to the park. Park in the first parking lot, immediately past the entrance booth. (If it's full, the farther lot will do). The Backbone Trail terminus[1] is at the far end of the first parking lot, next to the tennis courts. Facilities here include flush toilets, running water and picnic tables.

Next, drive your drop-off car (or rideshare) to the start of the hike. Head back down to Sunset Boulevard, return to the 405 freeway and take it north six miles to the 101 freeway. Go west on the 101 for about four miles and exit at Reseda Boulevard. Then head south on Reseda for about 3½ miles, following it steadily uphill, until it ends at a cul-de-sac in Marvin Braude Mulholland Gateway Park.

Park in in any of the hundreds of angled parking spaces lining Reseda Boulevard near the cul-de-sac. All-day parking within the park is fee-based, however you may park for free farther down the road outside the park boundary. (Look for the big line painted on the road separating the free spaces from the fee area.) The free spaces fill up quickly, so unless you're an early bird, expect an extra five-minute walk up the road from here. The parking lot has a pit toilet and a water fountain, the last reliable water source until the end of the hike.

**The hike:** For the first 2.6 miles, you will not actually be on the Backbone Trail but on a spur route that leads up from the parking area into Topanga State Park, eventually joining the Backbone Trail at The Hub. Virtually the entire stretch leading up to The Hub is on fire roads through chaparral, nearly always sun-exposed, and mostly uphill. The good news is (A) it's generally not all that steep and (B) after The Hub, the vast majority of your hike is heading downhill.

*Cathedral Rock*

Marvin Braude Mulholland Gateway Park, your starting point, is named for the long-serving L.A. City Councilman who fought for conservation of L.A.'s wild places and was a major hiking and bicycling advocate. The trailhead[2] is at the top of the cul-de-sac at the road's end. Begin by walking uphill on the fire road. (Don't take the footpath paralleling it to the left, which veers off into the developed section of the park.)

$1/10$ mile from the trailhead, leave the road for a single-track trail which angles off to the right.[3] This route, known as the Reseda Spur Trail, reconnects to the fire road in .2 miles but it's gentler, shorter, and far more pleasant than the road.

The short trail climbs steadily until it ends at dirt Mulholland Road[4] at .3 miles. Turn right on the road. (A left turn here would take you down into Caballero Canyon.)

You'll reach a "Y"-shaped junction of fire roads[5] at ½ mile. To the right would take you farther along Mulholland Road to Topanga Canyon Boulevard. Instead, head to the left on what is technically called "Temescal Ridge Trail, Fire Road #30." You'll walk through another gate, after which the fire road heads gradually downhill for a stretch. A sign notes that you're entering Topanga State Park.

As the fire road crosses a low saddle at 1.1 miles, it crosses a 4-way junction[6] where the Garapito Canyon Trail leaves to the right and the Bent Arrow Trail heads left. Continuing straight on the fire road from here, it's a 350 foot climb to reach The Hub – the longest single elevation gain on this hike.

The dirt road winds its way gently up the side of a ridge. Two miles into the hike, it catches up to the ridgetop and begins traversing the crest, providing some impressive views of Cathedral Rock, which looms straight ahead, resembling a cresting wave of tan sandstone.

Continue climbing until 2.6 miles, where you'll meet the Backbone Trail at The Hub,[7] a four-way intersection of dusty fire roads atop a sunbaked plateau. If from the name you were expecting hustle-and-bustle, a snack bar, and tourist amenities, you will be sadly disappointed. However, The Hub does offer more "civilization" than most points on the Backbone Trail in the form of a pit toilet, recycling and waste bins, and a kiosk with a sitting bench. There are no picnic tables or water facilities. At any rate, it's a good place to take a rest, and there almost always seem to be mountain bikers and hikers hanging out here chatting, comparing gear, or waiting for stragglers.

From The Hub, your route eastbound on Segment 1 of the Backbone Trail to Will Rogers Park leads straight ahead, while two other fire roads lead off to the right. The higher of the two roads (sharpest to your right) is the westbound Backbone Trail heading to Trippet Ranch. The lower road (Eagle Springs Fire Road) also takes you to Trippet Ranch, but via a slightly easier and far less scenic alternate route. (See Segment 2 for a description of both routes.)

Leaving The Hub, continue straight ahead on the Backbone Trail. You'll pass a sign that reads TEMESCAL RIDGE TRAIL TO CONFERENCE GROUNDS 7.7 KM (even though you're not ultimately heading to the church conference grounds in Temescal Gateway Park). At 2.7 miles, you'll pass Cathedral Rock on the left, the same monolith you saw earlier while heading up the spur trail.

> **SIDE TRIP:** *A short scramble up the hill will take you inside the "cathedral," which is a little bowl ringed by rock walls resembling a small fort. It's interesting but not particularly spectacular. The back side of the "cathedral" offers great views looking far down into Rustic Canyon.*
>
> *To get there, continue past the first side trail, which heads up before the rock. About a minute later, take the next path to your left,[8] and once at the top, turn left again.*

About a minute beyond Cathedral Rock, angle left at a false junction[9] where what appears to be a side road leaves off to the right. In fact, it goes nowhere and terminates after a few hundred feet of roadbed.

The Backbone Trail continues to climb very gently along a dirt road until 2.9 miles, where it heads gently downhill and crosses a sort-of saddle, offering lovely views in both directions. Looking left from the sharp drop-off, you can see the upcoming stretch of the Backbone Trail as it rounds the slope of a mountain. Below it you can also see the Bay Tree Trail heading more steeply down into a canyon.

Past the saddle, the Backbone Trail resumes its gentle climb along the sun-exposed fire road through chaparral. 3.1 miles into the hike, a disused, washed-out footpath[10] heads off to the left. This used to be the start of the old Rogers Road along the rugged north side of Temescal Peak, but several years ago it collapsed in a landslide. Some erroneous maps still list it as the Backbone Trail, but what's left of the old road is officially closed and is now reverting back to nature.

Continue straight ahead on the fire road/Backbone Trail until 3.2 miles, where you will finally leave the fire roads behind for some beautiful single-track trail. Take the footpath that heads off to the left[11] by a sign that reads MULTI-PURPOSE TRAIL and BACKBONE TRAIL TO WILL ROGERS STATE HISTORIC PARK 6.5 MILES.

Finally you are on a "real" trail, not something that was ever a road – in fact it was specifically built for the Backbone Trail to bypass the collapsed section of Rogers Road. During this .6-mile stretch, an overhanging mix of scrub oak, toyon, laurel sumac and ceanothus will offer you limited shade. In the spring, the area is brightened by all kinds of wildflowers.

The trail skirts just below the top of Temescal Peak, offering your first views of the skyscrapers of Century City and downtown L.A. At 3.3 miles, a small path crosses the Backbone Trail[12]. While the downhill path to the right rejoins the Temescal Ridge Fire Road, the uphill path to the left will take you to the top of Temescal Peak.

**SIDE TRIP:** *A short climb of about 2½ minutes will take you to the 2,126' summit of Temescal Peak, the highest point in the Santa Monica Mountains east of Topanga Canyon. From its arid, chamise-dotted top, you'll enjoy a superb 360° panorama, one of the better views in the entire mountain range.*

*Looking north you can see most of the San Fernando Valley. Turning eastward the view includes the high peaks of the San Gabriel Mountains including Mount Baldy, then Mount San Gorgonio and the San Bernardino Mountains, then downtown Los Angeles and Century City. To the south, the Palos Verdes Peninsula gives way to the ocean and Catalina Island. A turn to the west reveals Saddle Peak topped with communications towers, then distant Boney Mountain, and finally the impressive promontory of Eagle Rock.*

The Backbone Trail begins a moderate descent along the slope of Upper Temescal Canyon, at times carved out of the side of the steep mountainside, offering some great views looking down into the canyon. You'll meander in and out of spotty bits of shade through patches of elfin forest and chaparral. On clear days you can see the towers of Century City and, farther beyond, downtown Los Angeles.

At 3.8 miles, the single-track trail emerges from the chaparral and rejoins Rogers Road,[13] turning right to cross a saddle. The collapsed remnants of the old section of road head uphill to the left.

Just a few steps farther along the Backbone Trail, the Bay Tree Trail heads off to the left[14] and switchbacks steeply down into Upper Rustic Canyon. A sign at the trail entrance reads NOT A THROUGH TRAIL, because the various routes which exit from the other end of the Bay Tree Trail are all unmaintained and of questionable passableness.

For most of the remainder of this hike, you will be walking on the remnants of Rogers Road, an old bridle route designed and used by Will Rogers and his family when riding on his ranch nearly a hundred years ago. At times the old road is visible; at other points nature has reclaimed all of it except the single-track footpath of the Backbone Trail.

From the saddle, the trail switches over to favor the other side of the ridge, now offering views down into Rustic Canyon on your left. For the next half hour the trail doesn't change much; it stays mostly flat, occasionally working its way gently uphill towards a local high point, with just enough vegetation to frequently block out views but not enough to provide any sufficient shade. As you

*Lone Oak, high on the rim of Rustic Canyon, makes a welcome rest spot*

progress, Rustic Canyon on your left continues to deepen. You'll get occasional glimpses of L.A.'s Westside beyond the mountains.

At 4.7 miles, after a few minutes of moderate incline, you'll come to a right-turn "elbow" in the trail – particularly obvious on the map. The trail cuts only a small track through the very wide, overgrown roadbed. Views switch to your right side again.

About three minutes later, you will cross over a local high point of about 2,000 feet and begin a long, rocky, dry and sunny descent of about 400 feet. Expect no shade at all for the next half hour as what's left of the road narrows considerably and starts dropping down along the crest of the ridgeline, high on the eastern rim of Upper Temescal Canyon. This is arid country as you descend the south-facing slope past waist-high coastal sage scrub that looks perpetually thirsty: yucca, dodder, chamise, sage, and the occasional laurel sumac. Through this stretch you'll get your first decent views of the Westside. You can see Rustic Canyon dropping far below and beyond it, the Oz-like towers of Wilshire Boulevard and Century City.

It's not until 5.1 miles when you'll start really descending in earnest, heading down a steep section full of rocks and dusty gravel. You'll definitely want the proper footwear here.

The long downhill push lasts about five minutes until things start to flatten out. Soon after, the trail leaves the ridgeline and angles into Temescal Canyon, obscuring any further views of the city. Its proximity to the second largest metropolis in the country makes this quiet, lovely canyon seem all the more pristine and secluded.

After several minutes of hiking with no major landmarks, at about 5.6 miles you'll spy the Green Peak Communications Tower on your right, high on Temescal Ridge across the canyon. This small broadcasting facility is nowhere near as impressive as the name suggests, but at this point we'll take what landmarks we can get, and it will be the only one for the next half hour.

The trail dips tentatively down into the canyon. As it nears the bottom, a little more vegetation appears – thicker grass and shrubs, but still nothing to provide any reasonable shade.

5.8 miles into the hike, you will begin the first notable uphill stretch that you have encountered for quite some time. During this five-minute moderate climb, you may enjoy a few brief moments of tree shade. Along the way, you'll pass a small side path[15] to the right at 5.9 miles, which leads out to an unimpressive viewpoint.

After a brief break from the climb, a second five-minute uphill stretch tops a local high point at 6.4 miles, where another minor path[16] heads off to the right and goes nowhere in particular.

By now, the hike through the canyon may be getting a bit monotonous, and you might feel like you're stuck in one of those cheap Flintstones cartoons where the same background scenery repeats every few seconds. Take heart – a dramatic change is imminent. At 6.7 miles, after the trail descends to cross a small pass back to the western rim of Rustic Canyon, which has now deepened to 1,000 feet, you'll enter an increasingly shady stretch of overhanging oaks and tree-sized toyon. Gone is the mix of low chaparral and hot sun. Here the trail is soft and earthy, and the woods soon become so lush as to nearly resemble a jungle, including reeds, ferns and tall grass. It's as if you have stepped through a portal onto an entirely different trail.

In about ten minutes, the trail reaches a landmark at 7.1 miles known as Lone Oak. It's a huge old four-trunked coast live oak perched just to your left on the rim of the canyon, offering great views and one of the most attractive respites of shade on the whole trail. One could not think of a more appealing rest spot.

Immediately past the Lone Oak, the old Rogers Ridge Trail heads to the left while the Backbone Trail angles down and to the right.[17] (On a map the Rogers Ridge Trail might seem a viable alternate route as it parallels the Backbone Trail, tightly hugging the crest of the ridge before rejoining the BBT about a mile later. But watch out. While the Rogers Ridge Trail does *technically* go through, we don't recommend it. It starts off tamely enough, but the interior of it gets very nasty indeed. Portions are badly overgrown, others are extremely steep, rocky and slippery, with some badly rutted sections and serious slipping hazards. This poorly maintained trail is not for the casual hiker, the faint-of-heart, or the unsure footed, and it provides little in the way of views to justify the risk.)

Continuing southbound from Lone Oak, the Backbone Trail descends gently but steadily into the upper reaches of Rivas Canyon. After about seven or eight minutes, your descent steepens and the trail grows rockier and a bit rutted. Patchy bits of shade increase as you approach the canyon stream, which appears briefly on your right, supporting plenty of oaks and riparian woods. It's a short but particularly beautiful stretch of this hike.

Soon after (7.6 miles), the trail exits the riparian woods for good, as Rivas Canyon quickly drops far below the trail, leaving you walking high on its eastern slope. As you progress steadily downhill through a shaggy mix of grass, sage, ceanothus and laurel sumac, you'll begin to get some very impressive views ahead of Santa Monica, with its hotel towers lining the beach looking like toys from 1,100 feet above. Rogers Ridge on the left blocks any views of Los Angeles.

The trail passes directly beneath an impressive rock cut at 7.89 miles. The precarious old Rogers Ridge Trail is almost directly overhead at this point, running up and down along the crumbly spine of the ridge 150 feet above. As the

Backbone Trail makes a sharp left around the cut, wider views open up of Santa Monica, Pacific Palisades and the ocean down below the mouth of Rivas Canyon. You'll descend steeply for a minute or so, then pass under a smaller rock cut on the left.

*Santa Monica glimmers 1,100 feet below*

The trail begins a gentle climb, and after rounding a curve to the left at 8.1 miles, you'll get your first view of Chicken Ridge Bridge up ahead on the trail, as well as more terrific views of the Santa Monica coastline. After the viewpoint, the Backbone Trail continues its gentle climb up the ridgeline through a fairly shady section of overhanging ceanothus.

Soon after, at 8.2 miles, you'll reach a four-way junction.[18] The Rogers Ridge Trail rejoins the Backbone Trail here, coming down from the sharp left. The Josepho Spur Trail heads steeply downhill to the left, past a sign reading NO HORSES ON TRAIL.

A popular loop hike combines the Backbone Trail from Will Rogers Park with a return via the Josepho Spur Trail and the Rustic Canyon Trail. This loop is described at the end of this chapter (see "A Loop Alternative"), but if you've been hiking for several miles from Reseda Boulevard, we don't recommend using Rustic Canyon as an alternate route to Will Rogers Park. A quarter-mile trailless section through the narrows of Lower Rustic Canyon requires some class II scrambling over boulders and tramping through water in the year-round stream. This is no way to end a long trek over the mountains... unless you're a real glutton for punishment.

From the junction, follow the Backbone Trail to the right as it continues southeast along the ridge. You'll reach the local high point on this ridgetop section of trail at 8.4 miles, where a geodetic survey marker placed into the ground marks the summit of the 1,230-foot knoll.

From here onward, the trail takes on a different character. You're much closer to the city now, and the nearer you get to Will Rogers Park, the less serious the trail users seem to be. There may be people jogging with earbuds, some "hikers" wearing flipflops or spandex pants, people yakking on phones, kids running around, more dogs, bikes and distant city noise. The trail becomes wider and more heavily trodden.

From the survey marker, the trail starts descending fairly steeply. You'll get a panoramic view of most of Los Angeles from here, including miles of Pacific Ocean. In a few minutes, as you approach Chicken Ridge Bridge, a sign warns BICYCLISTS: TRAIL NARROWS, DISMOUNT AND WALK BEYOND BRIDGE. Just past the sign, the trail drops rapidly towards the bridge. Railings protect hikers and bicyclists from the steep drop-off.

You'll cross Chicken Ridge Bridge at 8.5 miles. It's not really a bridge per se, more like a boardwalk along a knife-edge, but Chicken Ridge is aptly named because it drops off sharply on both sides into two different canyons. Before the bridge was put in, this was a creepy stretch of trail indeed, especially if you travelled it on a horse as Will Rogers did, but now the bridge has taken much of the "chicken" out of the crossing.

*Descending toward Chicken Ridge Bridge*

After leaving the bridge, the Backbone Trail drops steadily along with the ridgeline. The medium height chaparral provides no shade – but oh the views! Amazing vistas abound in nearly every direction. On a clear day you can see as far as Mount San Jacinto to the east and Mount Laguna in San Diego County to the south, plus the full sweep of Santa Monica Bay and its famed "pearl necklace" of beaches. The entire city seems laid out before you, with the homes of three million people stretching to the horizon, punctuated by patches of skyscrapers. If you look to the left across Rustic Canyon, at times you can make out the thin vertical lines of the Murphy Stairs – a pair of insanely long old staircases dropping hundreds of feet down the opposite side of the canyon – the longest staircases in Los Angeles.

A little shortcut path heading to the left at 8.5 miles[19] and rejoining the Backbone Trail about .2 miles below,[20] is closed for plant rehabilitation. Please don't use it.

> **SIDE TRIP:** *At 8.8 miles, the Backbone Trail angles down and to the left, while a short side trail[21] heads straight ahead for about a minute's walk to a superb viewpoint. From here you can see… well… practically everything: the village of Pacific Palisades, the whole coastline curving from Santa Monica down to Palos Verdes, Catalina Island, the towers of Century City with downtown Los Angeles beyond, and farther to the left, Hollywood. You can see Rustic Canyon and its crazy staircases, and looking behind you get an expansive view of the rugged mountains over which you have just trekked.*

A few minutes further down the Backbone Trail, another side trail heads off to the right[22] in in exactly the same fashion, but this one doesn't provide nearly as good a viewpoint and is not recommended.

After the overlook side trail, the Backbone Trail drops steeply and soon you'll be able to look *down* on Inspiration Point, the flat-topped peak ahead that usually has a small crowd of view gazers atop it.

The trail then lessens its descent to a gentle pace and the chaparral thickens to provide an occasional hint of shade. It's at about 9.1 miles where you'll hit a switchback with a sharp left turn taking you down off of Rogers Ridge for good. A sign here guides you to the left, avoiding a disused trail straight ahead.

Then, 9.3 miles into the hike, you'll reach a major intersection[23] with a large old dirt bridle path known as the Inspiration Point Loop Trail, just below the back of Inspiration Point. You've just left Topanga State Park and crossed into the much-smaller Will Rogers State Historic Park.

From the junction, the Backbone Trail continues straight across the bridle path onto a smaller single-track path known as the Upper Betty Rogers Trail, named for Will Rogers' wife (Betty that is, not Upper Betty).

*SIDE TRIP: A brief diversion here will take you to the top of Inspiration Point, one of the most popular overlooks in all of Southern California. The easy side trip is well worth it. This aptly named, flat-topped promontory may not provide views of anything you haven't already seen today, but sure seems to pack it all in at one convenient spot, offering a 360° panorama. Benches, picnic tables and space to spread out make this an ideal rest spot. This is no remote lookout; here you are right on top of the city, with the faint sound of millions teeming down below, and you'll almost certainly have company.*

*This side trip is a mere .15 miles one-way with 50 feet of elevation gain. Turn right onto the Inspiration Point Loop Trail, and then in just .05 miles turn left onto the spur dirt road[24] leading up to the point.*

*ALTERNATE ROUTE: If you prefer a gentler but longer (and somewhat less scenic) route to the parking lot, consider finishing your trek by descending via the Inspiration Point Loop Trail. Being a loop, either direction will take you to the bottom.*

*A RIGHT TURN will take you steadily down the mountain to the Backbone Trailhead parking lot in 1.2 miles, offering plenty of views of the park and the ocean. Expect lots of company and no shade on the heavily travelled route. Along the way, the Backbone Trail crosses and re-crosses your route a few times, but simply stay on the bridle path all the way down, crossing above Will's house to the right in about .8 miles, then passing the historic stables on the left shortly thereafter.*

*Just beyond the stables (.95 miles from where you left the Backbone Trail), make a right on the paved service road[25] lined with a white rail fence and eucalyptus trees. (Will Rogers' house, just across the lawn, is an interesting spot to visit if you aren't too tired or grimy). Follow the service road for another .15 miles to its junction with the main park road[26] directly across from the polo field. The Backbone Trailhead[1] is 1/10 mile down the road to the right, just past the tennis courts.*

*ALTERNATE ROUTE: A LEFT TURN onto the Inspiration Point Loop Trail will return you to the parking lot in 1.1 miles via the slightly gentler eastern half of the loop, offering views of the equestrian area and polo field. Expect a bit more shade this way as the lower half of the old bridle path is lined with hundreds of stately eucalyptus trees.*

*After turning left off of the Backbone Trail, follow the bridle path loop gently down the ridge. Continue straight at a junction with the Bone Canyon Trail[27] on the right (¹/₁₀ mile from where you departed the Backbone Trail), which heads steeply down to the park corral. At ½ mile, take either fork at a "Y" junction[28] – both routes rejoin about a minute later.*

*At the bottom of your descent, the trail passes the old carpenter shop on the right where the Bone Canyon Trail rejoins it.[29] A few seconds later, at .85 miles, you'll meet up with the western half of the Inspiration Point Loop Trail at a junction where a paved service road heads off to the left.[25]*

*Turn left onto the service road, which is arrow-straight, lined with a white rail fence, and lies just across the lawn from Will's house. Follow it for another .15 miles to its junction with the main park road[26] directly across from the polo field. From here the Backbone Trailhead[1] is ¹/₁₀ mile down the road to the right, just past the tennis courts.*

The Upper Betty Rogers section of the Backbone Trail is shady, cool, scenic, free of bicycles, and less popular than either of the bridle path alternate routes. From the junction, the single-track Backbone Trail winds gently along the southern slopes below Inspiration Point through tall chaparral, offering intermittent views of the park and the ocean.

A side trail with steps leading off to the right[30] at 9.6 miles goes back up to Inspiration Point via the hard way. The route is very steep, gravelly, heavily rutted, and makes a lousy way to get up to the point when compared to the main access.

For the next few minutes, when lower chaparral permits, the Backbone Trail offers fine views of the polo field down below ringed by verdant plateaus. A few minutes later, at 9.8 miles, it rejoins the same bridle path that you crossed earlier.[31] Turn left, heading down the bridle path.

After descending the bridle path for four or five minutes, the Backbone Trail branches off of it onto another single-track trail to the left[32] at mile ten. This short footpath, known also as the Lower Betty Rogers Trail, is shorter but much steeper than continuing down the bridle path, as it drops straight down to the parking lot with no fooling around, descending a dry, rocky slope through plenty of yucca and sage.

After about a minute of steep descent on the rutted path, you'll have a clear view of the parking lot off to your right. You're almost done. Hang in there!

The steep trail crosses the bridle path[33] again at 10.1 miles. It drops for a few more minutes via occasional steps, crosses a tiny bridge over a ditch and passes the tennis courts. And then, after hiking 10.2 miles… drumroll… you reach the parking lot and the eastern terminus of the Backbone Trail.[1]

# A SHORTER ALTERNATIVE

Don't feel like hiking ten miles? No problem! Many of the best features of this segment lie relatively close to the eastern trailhead and can be visited via a four-mile out-and-back hike up the Backbone Trail from Will Rogers State Historic Park to the Josepho Trail junction. On this trip you'll enjoy unbeatable ocean and city views, quiet woods, Chicken Ridge Bridge, Inspiration Point, and even a visit to Will Rogers' home. Add another 2¼ miles if you extend the out-and-back trip up to the quiet and peaceful lookout of Lone Oak for lunch.

# A LOOP ALTERNATIVE

Another way to see the Will Rogers Backbone area is via the popular Rustic Canyon/Backbone loop hike. If you don't mind some annoying graffiti and a quarter mile of off-trail trudging up a wet streambed with a little class II boulder scrambling, you could follow this loop as it heads up Rustic Canyon, then climbs about 650 feet up the Josepho Trail, then returns to Will Rogers Park via the Backbone Trail. Along the way you'll experience the wild and challenging Rustic Canyon Narrows, the ruins of Murphy Ranch (with its history of WWII Nazi sympathizers who used the buildings for secret broadcasts), a visit to the Murphy Stairs, and a spectacular finish descending along the Backbone Trail via Chicken Ridge Bridge, Inspiration Point, and Will Rogers' home. Not bad for a loop trip of about five miles.

While this loop hike is reasonably popular, it's lost much of its charm as of late. Why? Despoliation at the hands of graffiti "artists." For years the abandoned buildings of Murphy Ranch have been covered with graffiti, but recently the tagging has spread across much of the surrounding canyon area to the sycamores, the boulders, and other historic sites – nothing "artistic" about that.

Before you get to Murphy Ranch, you'll have another challenge: a quarter-mile trailless section through the narrows of Rustic Canyon, requiring some class II scrambling over boulders and small ledges, plus tramping through water in the year-round stream. To some this can be a fun adventure; to others it's a bummer.

### RUSTIC CANYON/BACKBONE LOOP TRAIL DETAILS

To take the loop, start at the Backbone trailhead[1] by the parking lot. Follow the paved main park road eastbound as it skirts the northern edge of the polo field. Pick up the Rustic Canyon Trail at the far end of the polo field[34] (.2 miles) and follow it as it steadily winds downhill into Rustic Canyon.

At .7 miles, the trail reaches the stream at the bottom of the canyon... and promptly disappears.[35] The hike heads north up-creek through the rocky narrows. For the next ¼ mile, the barely existent footpath winds in and out of the streambed and crosses it several times, requiring a choice between either walking in water or scrambling over ledges, making this section very slow and rugged – not suitable for everyone.

*(continued on page 58)*

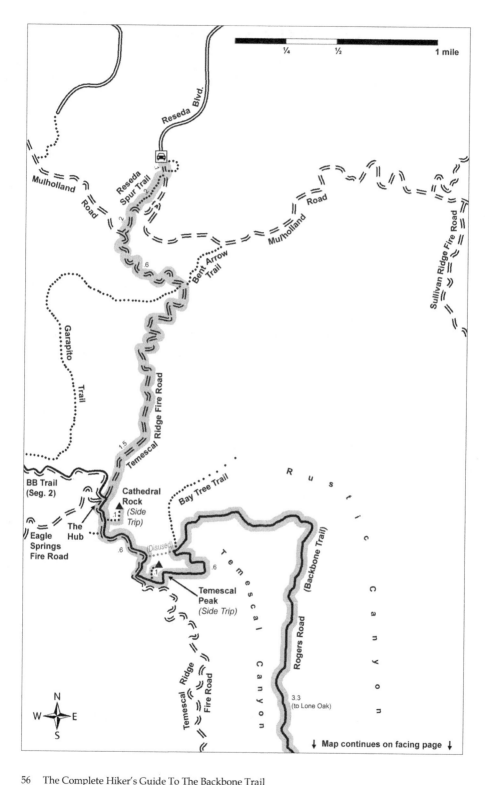

¼  ½  1 mile

Reseda Blvd.

Mulholland Road

Reseda Spur Trail

Mulholland Road

Sullivan Ridge Fire Road

Bent Arrow Trail

.6

Garapito Trail

1.5

Temescal Ridge Fire Road

BB Trail (Seg. 2)

Cathedral Rock *(Side Trip)*

.1

Bay Tree Trail

R u s t i c

The Hub

Eagle Springs Fire Road

.6

(Disused)

.6

(Backbone Trail)

T e m e s c a l

.1

Temescal Peak *(Side Trip)*

C a n y o n

Rogers Road

Temescal Ridge Fire Road

3.3 (to Lone Oak)

C a n y o n

N
W    E
S

↓ Map continues on facing page ↓

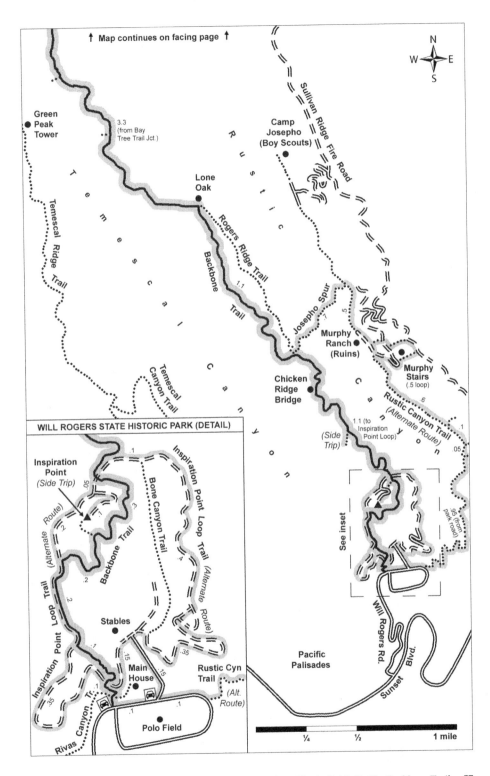

↑ Map continues on facing page ↑

N
W — E
S

Green
Peak
Tower

3.3
(from Bay
Tree Trail Jct.)

Camp
Josepho
(Boy Scouts)

Sullivan Ridge Fire Road

R
u
s
t
i
c

Lone
Oak

Temescal Ridge Trail

T
e
m
e
s
c
a
l

Rogers Ridge Trail

Backbone Trail

1.1

Josepho Spur

.7

.5

Murphy
Ranch
(Ruins)

Murphy
Stairs
(.5 loop)

Temescal Canyon Trail

C
a
n
y
o
n

Chicken
Ridge
Bridge

Rustic Canyon Trail
(Alternate Route)

.6

.1

C
a
n
y
o
n

.05

1.1 (to
Inspiration
Point Loop)

(Side
Trip)

(Side
Trip)

See inset

.95 (from
park road)

WILL ROGERS STATE HISTORIC PARK (DETAIL)

Inspiration
Point
(Side Trip)

.1

Inspiration Point Loop Trail

.05

.3

Bone Canyon Trail

.3

.7

(Alternate Route)

.2

Backbone Trail

.2

Inspiration Point Loop Trail

.2

.4

Stables

.1

.15

.35

(Alternate Route)

.15

Main
House

.15

Rustic Cyn
Trail

.35

Inspiration Point Loop Trail

Rivas Canyon Tr.

.1

Polo Field

.1

(Alt.
Route)

Pacific
Palisades

Will Rogers Rd.

Sunset Blvd.

¼        ½         1 mile

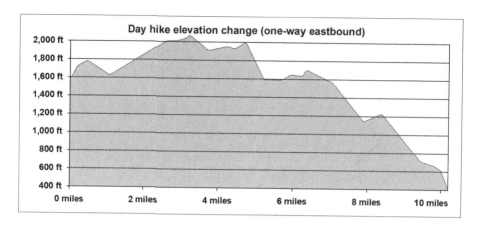

Day hike elevation change (one-way eastbound)

In the narrows, expect a good possibility of getting your feet wet, since the spring-fed stream contains water year-round. If you wore convertible hiking pants, it's advised to keep the long legs on during this section, since this is a prime habitat for poison oak.

After several minutes of streambed stomping, the faint trail will begin to make itself discernible just to the right (east) of the stream at roughly .9 miles.[36] Follow it, crossing the stream several more times, until you get to a minor junction at 1.15 miles.[37] Take the path to the left that crosses the stream again, then climbs briefly uphill away from the water. (If you overshoot the junction, you'll quickly notice the trail veer sharply to the right, heading steeply up some railroad-tie steps; go back.)

From here on, you'll have a somewhat more legit trail, although it is officially unmaintained. Follow the trail up-canyon above an old dam at 1.2 miles. Head down to a little junction[38] at the crest of the dam, then follow the main trail left up the canyon.

The trail continues to improve as you emerge from the narrows through lush, jungle-like vegetation of tangled vines, reeds, wild asparagus, ferns and poison oak. Veer left at a tricky junction[39] (1.3 miles) where a side path heads uphill to a ravine.

At 1.6 miles, a small path[40] to the right provides access to the ruins of Will Rogers' cabin. Only the cement foundation, fireplace and chimney remain, all graffiti-covered.

As you continue, the canyon widens into a large, sheltered floodplain that supports a grassy oak and sycamore savannah. You're entering a section once known as Murphy Ranch, built as a self-sufficient compound by a wealthy couple and used (perhaps without their knowledge) by Nazi sympathizers to beam short-wave broadcasts to Germany during World War II. The mystique of this place has, alas, attracted many graffiti taggers who have marred the area with paint and litter.

Several small paths crisscross the Murphy Ranch area; when in doubt, follow the main path that keeps relatively close to the stream. Continue past the ruined power generator with its rusting diesel tank (1.65 miles) and, a couple of minutes later, past the smashed remains of the sheet metal machine shop.

At 1.9 miles, the trail joins up with an old semi-paved road[41] and follows it via a gentle left. (A two-minute detour up this road to the right will take you to a small footpath[42] leading to the base of the Murphy Stairs, the first of two mind-bogglingly long staircases, each of roughly 500 steps, both of which climb three hundred feet up the canyon slope to Sullivan Fire Road. If you're into creepy, semi-abandoned staircases and are up for a workout, you could make a loop of it, climbing these steps to the paved fire road[43], then walking west (left) about two minutes on the road, and then descending the other staircase.[44] Note that the second staircase jogs right[45] for a minute or two at a semi-paved road about ¾ of the way down, then continues down the hill[46] to join the Rustic Canyon Trail[47] at the ruined power station mentioned next.)

To continue along the Rustic Canyon Trail, follow the crumbling road north, passing the shell of an old power station off to your right at mile two. This was once the broadcasting point for the aforementioned Nazi sympathizers. It's been covered top to bottom with graffiti for years.

The road continues past various other ruins, concrete foundations, overgrown chimneys and steps leading to nowhere. When the road ends at 2.1 miles, your hike continues straight as a single-track trail. From here you'll leave the Murphy Ranch ruins behind for more natural surroundings.

2.4 miles into the hike, you'll come to a junction with the Josepho Spur Trail.[48] For many years, an old wooden barn here at the junction made for a picturesque surprise, but it has recently dilapidated and collapsed. Take the Josepho Trail to the left.

The trail immediately crosses Rustic Stream, then begins to work its way up the western slope of the canyon. During the next ⅔ mile, the mostly steep trail climbs a relentless 650 feet through a semi-shady mix of oak, manzanita, laurel sumac, ceanothus and toyon, until it joins the Backbone Trail[18] at the top of the ridge at 3.1 miles.

From here, pick up the Backbone Trail southbound as it continues for another two miles down the ridgeline, passing Chicken Ridge Bridge, Inspiration Point, Upper and Lower Betty Rogers Trails and Will Rogers' house, then back to the parking lot in Will Rogers Park. (See BBT description in the earlier part of this chapter.)

---

### GPS COORDINATES

| | | |
|---|---|---|
| 1. N34° 3.259', W118° 30.803' | 2. N34° 7.880', W118° 33.190' | 3. N34° 7.828', W118° 33.181' |
| 4. N34° 7.694', W118° 33.327' | 5. N34° 7.591', W118° 33.401' | 6. N34° 7.375', W118° 33.082' |
| 7. N34° 6.473', W118° 33.490' | 8. N34° 6.369', W118° 33.491' | 9. N34° 6.339', W118° 33.503' |
| 10. N34° 6.213', W118° 33.254' | 11. N34° 6.128', W118° 33.307' | 12. N34° 6.104', W118° 33.203' |
| 13. N34° 6.259', W118° 33.139' | 14. N34° 6.277', W118° 33.123' | 15. N34° 5.402', W118° 32.488' |
| 16. N34° 5.133', W118° 32.245' | 17. N34° 4.837', W118° 31.786' | 18. N34° 4.205', W118° 31.289' |
| 19. N34° 4.009', W118° 31.139' | 20. N34° 3.981', W118° 31.075' | 21. N34° 3.894', W118° 31.007' |
| 22. N34° 3.822', W118° 30.954' | 23. N34° 3.646', W118° 30.819' | 24. N34° 3.619', W118° 30.825' |
| 25. N34° 3.382', W118° 30.750' | 26. N34° 3.276', W118° 30.701' | 27. N34° 3.670', W118° 30.733' |
| 28. N34° 3.428', W118° 30.628' | 29. N34° 3.384', W118° 30.720' | 30. N34° 3.549', W118° 30.830' |
| 31. N34° 3.511', W118° 30.893' | 32. N34° 3.374', W118° 30.867' | 33. N34° 3.307', W118° 30.810' |
| 34. N34° 3.266', W118° 30.591' | 35. N34° 3.396', W118° 30.394' | 36. N34° 3.589', W118° 30.388' |
| 37. N34° 3.765', W118° 30.384' | 38. N34° 3.810', W118° 30.391' | 39. N34° 3.875', W118° 30.435' |
| 40. N34° 4.000', W118° 30.619' | 41. N34° 4.187', W118° 30.837' | 42. N34° 4.164', W118° 30.728' |
| 43. N34° 4.213', W118° 30.627' | 44. N34° 4.249', W118° 30.660' | 45. N34° 4.206', W118° 30.774' |
| 46. N34° 4.231', W118° 30.825' | 47. N34° 4.211', W118° 30.872' | 48. N34° 4.470', W118° 30.980' |

# THE HUB TO TRIPPET RANCH
## ("MUSCH TRAIL" & "EAGLE ROCK FIRE ROAD")

**BBT segment length: 3.5 miles**
**Day hike length: 3.5 miles one-way**
**Suggested day hike: west to east, out-and-back**

**Elevation gain westbound: 248 feet**
**Elevation gain eastbound: 1,068 feet**
**Difficulty: moderate – Shade factor: 25%**

**Bikes not allowed on single-track section**
**Horses allowed – Dogs not allowed**

**Access: N34° 5.613', W118° 35.249'**

This wonderfully diverse hike takes you across rolling meadows, through beautiful patches of riparian woodlands, past sweeping views of the Topanga Canyon area and the San Fernando Valley, and can include a great side trip up spectacular Eagle Rock.

The first half of the hike follows a single-track trail (also known as The Musch Trail), which crosses the grassy meadows of Trippet Ranch, something very rare in the Santa Monica Mountains. It dips down into a couple of lush canyons with deep woods and heads along a sandy shelf offering great views of the upper Topanga Basin. The trail then quits its pleasant meandering and heads up a canyon in earnest, climbing a long uphill stretch before converging with a fire road.

*Entrance to Trippet Ranch*

The heavily sun-exposed fire road takes you near Eagle Rock, with its sheer drop-off, honeycombed mini-caves and high ledges, before working its way up to The Hub, a lonely convergence of dirt roads on the plateau rooftop of Topanga State Park. With no car access at The Hub, you'll need to turn around and hike back 3½ miles or follow through on to adjacent Segment 1.

You may very well have company on the fire road portion, as this area is big with mountain bikers. But the first half (the Musch Trail) will leave you feeling like you've gone far out into the country, even though you are actually only a few miles from the Westside.

Ocean panorama from *Eagle Rock Fire Road*

**Directions to trailhead:** From Pacific Coast Highway, turn north onto Topanga Canyon Boulevard and follow it up through both the canyon and Topanga village for 4.7 miles, then turn right on Entrada Road. (If you're coming from the 101 freeway, exit Topanga Canyon Boulevard and follow it south over the mountains for 7.7 miles, then turn left onto Entrada Road.)

Follow Entrada as it winds steadily up the hill, bearing left at every major junction. 1.1 miles after leaving Topanga Canyon Boulevard, you'll enter Topanga State Park in an area known as Trippet Ranch. Park in the large day-use lot at the end of the road (fee required). The area has plenty of facilities such as grills, picnic tables, restrooms, drinking fountains, a visitors center and a ranger station.

**The hike:** Locate the trailhead[1] at the northeast corner of the Trippet Ranch Parking Lot, to the right of the restrooms. As the name suggests, this place was once a working cattle ranch. It was bought by Judge Oscar A. Trippet and his wife Cora in 1917. Oscar's son (Oscar Jr.) further developed the property in the 1940s, adding a barn, a superintendent's house and a skeet lodge (now the visitors center). In the 1960s, with the property under threat of massive development, a citizens' group formed to preserve it and eventually convinced the state to acquire it.

The hike starts out on a small paved road heading past a sign that reads MUSCH TRAIL. (Note that a different route leaving the southeast corner of the lot via a dirt road is NOT the Backbone Trail, but is an alternate covered later in this chapter.)

Follow the road past the MUSCH TRAIL sign. It soon crosses a tiny bridge by a seasonal pond on the right, then arrives at a junction with the Backbone Trail[2] (although the signs don't label it as part of the BBT system). The route continuing straight ahead on the road is signed MUSCH TRAIL, while the single-track route to the left is the ominous-sounding DEAD HORSE TRAIL (see description for Segment 3). Both of these routes are also the Backbone Trail. Continue straight on the road.

You will immediately enter an open area known as Musch Meadows. This section is ideal for spotting deer as well as smaller critters like rabbits and squirrels.

At ¹⁄₁₀ mile, the Backbone/Musch Trail angles off to the right onto a signed single-track footpath across a beautiful meadow.[3] There's a little water spigot and drinking fountain at the junction.

After crossing the meadow, you'll enter the first of several patches of riparian woodlands and come to a wooden footbridge over a seasonal stream (.2 miles).

For the next half mile your hike will alternate between chaparral (chamise, laurel sumac, manzanita and toyon) and riparian oak woodlands as it dips down into the occasional seasonal-stream ravine. You'll get better and better views looking west over Topanga Canyon. Parts of this trail can be rutted and rocky, so hiking boots are a necessity.

After a handful of these miniature stream

*Into the meadow*

crossings, at .5 miles you'll come to a minor junction[4] where an old rutted trail leads down some steps to your left, eventually becoming Hillside Drive in Topanga Village. The Backbone Trail continues straight here; a small sign facing the other way reads TRAIL for returning hikers.

During the next ¹⁄₃ mile, the trail follows a low sandstone shelf through sun-exposed chaparral, offering increasingly better views of the upper Topanga Basin and surrounding mountains. It climbs gently, passing just above a housing development which stays mostly hidden. A bench at .6 miles marked IN LOVING MEMORY OF ALAN SHERMAN makes a good rest spot with a fine view of the basin.

.8 miles into the hike, the Backbone Trail veers right at a junction.[5] A sign reads MUSCH MEADOW TRAIL. (The side trail to the left becomes Prier Road and descends to the housing community below.)

Within a minute, you'll enter a field where the Backbone Trail branches to the left,[6] leaving the old remnants of Prier Road. (A sign here directs you to the left.) Within another minute you'll enter Musch Trail Camp and cross the small paved campground access road[7] at .9 miles. The campground, designed for through-hikers and overnight equestrians, is tiny but picturesque, with flush toilets, clean sinks, water spigots, picnic tables, a small corral and a peaceful eucalyptus grove.

It doesn't take long to cross the campground, and within a minute you're leaving this tranquil place. Soon you'll be heading straight across a huge, rolling, grassy meadowland. In the spring, it's wonderfully lush and green, providing expansive views of nearly all of Topanga Canyon.

In the meadow, at mile one, you'll reach another junction.[8] Take the trail to the right, which is signed MUSCH TRAIL TO EAGLE JUNCTION. After a brief change of scenery through a copse of large oaks with ferns, the trail steeply climbs the sloping meadow until you reach a small saddle, which offers a 360° panorama of the meadow and mountains. An overgrown side trail[9] heading off to the right from here is closed for plant rehabilitation.

From the saddle, the Backbone Trail descends gently into a side canyon, leaving the meadow for chaparral. Then, at 1.4 miles, you'll enter a lovely wooded glade and cross a year-round stream over a big metal pipe. A few minutes later you'll repeat the process with another lovely glade and stream.

After crossing this second stream, the trail turns and begins a steady, moderately steep climb up the canyon's north slope toward Eagle Junction, gaining 380 feet in 2/3 of a mile, utilizing occasional steps along the way.

The trail climbs through semi-shady tall chaparral at first, which gradually lowers as you ascend, eventually providing no shade at all but allowing frequent views looking back on the upper part of Topanga Canyon.

After a while, large white sandstone outcrops begin to appear on the opposite side of the canyon, baby cousins of the bigger Eagle Rock formation which lies farther ahead along the hike. The climb increases from moderate to steep for the next few minutes. About a minute into the steep stretch, the route climbs to reach a minor "T"-shaped junction[10] at 1.7 miles, where the Backbone Trail turns to the right. The short path to the left goes nowhere in particular and quickly fades away.

*Picnic area at Musch Trail Camp*

The trail continues its steep uphill trudge until roughly 1.9 miles, where it lessens in severity and switchbacks around the head of a ravine; here some impressive views open up on the right looking back down the canyon toward the Topanga Basin. On a clear day you can see several mountain ridges through the gap above Topanga Canyon. You can also spot the fire road running along the ridgetop up ahead, which you will soon join.

2.1 miles into the hike, you'll reach the fire road, the ridgetop, and the end of the Musch Trail at a spot known as Eagle Junction.[11] Here two different fire roads make a "Y"-shaped intersection on your left, while a single fire road heads south along the ridge to your right. The Backbone Trail follows the left fork of the "Y"-shaped junction – the upper one – also known as Eagle Rock Fire Road.

**ALTERNATE ROUTE:** *If you're wiped out on your way back from an out-and-back hike and want to return to your car ASAP, a shortcut heading south on the Eagle Springs Fire Road will make your trip easier than taking the Musch Trail back. Instead of wiggling in and out of wooded valleys like the Musch Trail, the dirt road heads directly down the crest of the ridge until it dumps you out at the Trippet Ranch Parking Lot, 1.6 miles from Eagle Junction. This route chops off a good third of the mileage and some elevation gain, but sacrifices much of the scenic appeal of the Musch Trail.*

*From Eagle Junction, head southwest on Eagle Springs Fire Road (opposite the "Y" fork). You'll soon pass a sign that reads* EAGLE SPRINGS FIRE ROAD TO TRIPPET PARKING LOT. *The road initially heads uphill to a high point at .3 miles, then begins a steady descent along the rolling, grassy ridgetop.*

*About .8 miles into the route, where the grass turns to chaparral dotted with oaks, you'll begin to see the Trippet Ranch Parking Lot down below on the right. At mile one, pass the signed Santa Ynez Trail[12], which heads left down into Santa Ynez Canyon.*

*A few minutes later, a single-track footpath signed* NATURE TRAIL[13] *crosses the route. A right on this trail leads to the parking lot via the visitors center, but staying on the fire road is easier. Follow the road for another minute through a stand of coast live oaks, passing a dirt road to the left[14] and, a few steps farther, another single-track path to the right. Keep right at both of the next two road junctions,[15 & 16] before finally merging onto the visitors center access road[17] just steps from where it meets the parking lot.[18]*

At Eagle Junction, as the Backbone Trail transitions from the Musch Trail to Eagle Rock Fire Road, it abruptly changes character. Gone is the quiet single-track beauty of the Musch Trail with its wooded glades and rolling meadows. From here on, the Backbone Trail is wider and more arid, following an old fire road, which at times feels more like a trail. This route hugs the crest of a high ridge, skirting the perimeter of the 2021 Palisades Fire. It offers little shade except for winter mornings when parts of it are shadowed by overhanging bluffs, but it does offer a multitude of excellent views looking north to the San Fernando Valley and south to the ocean.

Follow the rocky road as it climbs steadily from Eagle Junction up the ridge. At the top (2½ miles), a well-trodden path signed EAGLE ROCK heads to the right.[19]

**SIDE TRIP:** *The short detour to visit Eagle Rock is well worth it, as it's the most spectacular rock formation in Topanga State Park and one of the most popular hiking destinations in the Santa Monica Mountains. This huge pink and tan sandstone monolith hanging over the edge of Santa Ynez Canyon is riddled with holes and caves, offering a great view from the top of its sheer southern face.*

*A minute's walk plus a short scramble will get you to the top of the rock, where you'll enjoy views looking south down the canyon, past the distant development of Palisades Highlands, and over the bay as far as Palos Verdes Peninsula.*

*A bench at an overlook just before the rock makes a perfect rest spot. From here,[20] a .05-mile side path heads south up the ridge to a sharp knoll looking down on Eagle Rock. The knoll's 360° panorama of the Topanga area may be the best view on the entire hike. The path does, however, become steep and gravelly, a potential slipping hazard.*

From Eagle Rock, the dirt road generally follows the ridgeline all the way to The Hub. Along the way, a few minor use paths diverge and then re-converge with the main route, but none of them offer any decent hiking alternatives, and some are dangerous.

You'll descend to a saddle at 2.6 miles where you'll pass Cheney Fire Road[21] and, just past it, the Garapito Canyon Trail,[22] both on your left.

As the BBT climbs the ridge, a brief spur on the right[23] at 2.9 miles leads up to a bench with a fine view looking west. (Resist the temptation to continue past the bench on the use path, which becomes rocky, very steep, and hazardous.) Soon after, you'll crest the hike's high point at mile three, where the roadbed has exposed a mini moonscape of conglomerate rock formations.

*Eagle Rock looms over Eagle Springs Fire Road*

**SIDE TRIP:** *Shortly before The Hub, a minute's walk on a side path[24] at 3.3 miles will take you up a knoll just north of the BBT. The top offers a 360° panorama of the area, including the sprawling San Fernando Valley and the hills beyond the Conejo Valley.*

Once past the side path, the fire road descends the shoulder of a ridge to meet The Hub[25] at 3.5 miles. This is the end of Segment 2 and a good turnaround point.

There's not a lot at The Hub, yet hikers and bicyclists always seem to be hanging out taking a break at this lonely confluence of dusty fire roads at about 2,000 feet. There's a kiosk with a bench, permanent pit toilets and trash cans, but no water.

**SIDE TRIP:** *From The Hub, a short side trip will take you up into the "cathedral" of Cathedral Rock, a little sandstone bowl ringed by rock walls resembling a small fort. It's only mildly interesting, but the back side offers great views looking far down into Rustic Canyon. To get there, turn right, staying on the road-like Backbone Trail. Follow the road for about two minutes, passing the first side path, which heads up just before the rock. About a minute later, take the next path to your left.[26] A brief scramble up the short but steep path, followed by another left, will take you inside the "cathedral."*

From The Hub you have several choices. You could continue on the Backbone Trail to Will Rogers State Historic Park by turning right (see Segment 1). If you were to make a 90° left, that would lead you down to the parking lot at the end of Reseda Boulevard. You could, of course, return the way you came via the Backbone Trail – or you could return to Eagle Junction via a slightly easier alternate route on Eagle Springs Fire Road, the dirt road downhill to the sharp right (see next page).

**ALTERNATE ROUTE:** *If you're turning around at The Hub, another way to return to Eagle Junction is via Eagle Springs Fire Road, the lower of the two fire roads leading there from here. This road will get you back to the junction with about the same distance and elevation gain as the Backbone Trail, but via a slightly easier and more gently graded route. It also provides some excellent up-close views looking up at Eagle Rock.*

*The downside is that it makes for a rather boring trip on a gravel road, wide and dusty, offering nary an ounce of shade as it cuts right through the 2021 Palisades Fire burn area, with the occasional mountain bike whizzing by.*

*From The Hub, take the fire road which leads downhill just to the left of Eagle Rock Fire Road. The road*

A public service from the Department of Pointless Signage

descends the mountain, offering views of Cathedral Rock to the left. Along the way, you'll cross over a seasonal stream, with an official-looking metal sign astutely labeling it CULVERT. Really, was this signage necessary? If we mistook it for a "gully" or a "gulch," how bad could that be?

Shortly beyond, as the road turns right, you'll get a brief view towards the sea and the red-tiled roofs of Palisades Highlands. Another minute further down the road (½ mile from The Hub), the fire road crosses another culvert (again inexplicably signed CULVERT).

As you continue steadily down to Eagle Spring for several minutes, you'll get impressive views of Eagle Rock across Santa Ynez Canyon.

At about a mile, the fire road reaches its low point, crossing just below Eagle Spring. The sycamores here are a telltale sign of a steady water source... and so is the huge metal pipe under the trail, from which splashing can often be heard. Just in case you didn't hear it, another sign labels this as – wait for it – CULVERT!

A few seconds later, you'll pass a tiny side trail[27] signed EAGLE SPRING on your right. (As a side trip, Eagle Spring isn't recommended. A short trudge up the overgrown path will take you past the crushed remains of the old wooden water tanks, and then to the underwhelming spring itself. Stay clear of the spring – a sign warns of unsafe drinking water, while heavy thickets of poison oak may pose a major hazard here.)

After Eagle Spring, it's a gentle climb to Eagle Junction.[11] Along the way you'll pass below Eagle Rock, with frequent vistas looking up at it. It's quite obvious which one of the sandstone outcrops is Eagle Rock; it's several times larger than any others, riddled with holes and caves, and you'll frequently see hikers sitting atop it. Birds – not necessarily eagles – seem to love circling it.

*Once past the rock, you'll get fine views of the coast, with the mesa-like flat top of Palos Verdes Peninsula visible far in the background. You'll rejoin the Backbone Trail at Eagle Junction[11] 1.3 miles from where you left The Hub.*

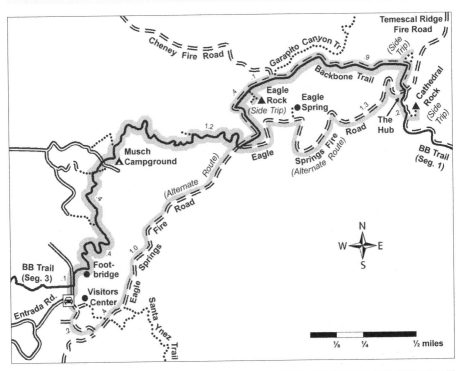

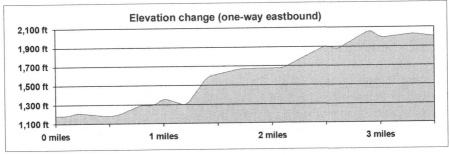

## GPS COORDINATES

| | | |
|---|---|---|
| 1. N34° 5.613', W118° 35.249' | 2. N34° 5.652', W118° 35.229' | 3. N34° 5.706', W118° 35.213' |
| 4. N34° 5.908', W118° 35.064' | 5. N34° 6.131', W118° 35.107' | 6. N34° 6.143', W118° 35.074' |
| 7. N34° 6.188', W118° 35.044' | 8. N34° 6.255', W118° 34.997' | 9. N34° 6.325', W118° 34.861' |
| 10. N34° 6.288', W118° 34.584' | 11. N34° 6.244', W118° 34.350' | 12. N34° 5.592', W118° 34.950' |
| 13. N34° 5.535', W118° 35.022' | 14. N34° 5.508', W118° 35.058' | 15. N34° 5.494', W118° 35.100' |
| 16. N34° 5.536', W118° 35.213' | 17. N34° 5.587', W118° 35.206' | 18. N34° 5.596', W118° 35.240' |
| 19. N34° 6.470', W118° 34.300' | 20. N34° 6.459', W118° 34.290' | 21. N34° 6.531', W118° 34.181' |
| 22. N34° 6.535', W118° 34.155' | 23. N34° 6.581', W118° 33.889' | 24. N34° 6.579', W118° 33.499' |
| 25. N34° 6.473', W118° 33.490' | 26. N34° 6.369', W118° 33.491' | 27. N34° 6.351', W118° 34.090' |

# SEGMENT 3:
# TRIPPET RANCH TO TOPANGA CANYON
## ("DEAD HORSE TRAIL")

**BBT segment length: 1.2 miles**
**Day hike length: 1.2 miles one-way**
**Suggested day hike: east to west, out-and-back**

**Elevation gain westbound: 63 feet**
**Elevation gain eastbound: 461 feet**
**Difficulty: easy to moderate – Shade factor: 45%**

**Bikes not allowed – Horses allowed – Dogs not allowed**

**East access: N34° 5.613', W118° 35.249'**     **West access: N34° 5.694', W118° 36.091'**

This segment is also known as the Dead Horse Trail, but don't let that name scare you off. The horse for which the trail is named must have been pretty sickly if it gave up the ghost on *this* trek, because it's a relatively easy hike. It doesn't boast any superlatives – no peaks, no waterfalls, no ocean views, no major destinations – and yet it's a wonderful outing and one of the most popular stretches of the Backbone Trail. Why? Because it's simply a peaceful, beautiful, and relatively easy ramble through many different types of Santa Monica Mountains environments. In not much more than an hour of round-trip walking, the path winds its way down into Topanga Canyon through serene patches of oak woods, across a bridge in a shady glen with croaking frogs, and through the rolling grass meadows of Trippet Ranch. More often than not we see deer on this hike, but no dead horses to date.

If you return via the short alternate route along the 92 Trail, you'll replace the open grasslands with a stretch of beautiful deep oak woods and encounter the rusting remains of a disused old floodgate and other relics from the ranch era.

Due to its nearby proximity to L.A., this segment is one of the busiest, so be prepared to come across many other trail users, a few of whom don't know how to behave on a hike. Some folks, we suspect mostly ones who've just popped out of the city for an hour or two, talk so loudly that they scare away the wildlife, which is a big part of the enjoyment of

*A common sight at Trippet Ranch*

hiking this trail. We encourage visitors here to keep their voices down and put the phone away so that everyone can enjoy the nature found along this gem of a hike.

**Directions to trailhead:** From Pacific Coast Highway, turn north onto Topanga Canyon Boulevard and follow it up through both the canyon and Topanga village for 4.7 miles, then turn right on Entrada Road. (If you're coming from the 101 freeway, exit onto Topanga Canyon Boulevard and follow it southbound up over the mountains for 7.7 miles, then turn left on Entrada Road.)

Take Entrada Road up the hill. Bypass the Dead Horse Parking Lot on the left and continue on Entrada, bearing left at every major junction. About 1.1 miles up Entrada after leaving Topanga Canyon Boulevard, you will reach the entrance to Topanga State Park in a woodsy area known as Trippet Ranch. Park in the day-use parking lot at the end of the road (a fee is required). The area

*Backbone Trail between the meadow and the woods*

has drinking water, flush toilets, picnic tables, an info kiosk and a ranger station.

To find the Backbone trailhead,[1] walk to the northeast corner of the parking lot (to the right of the restrooms), where a small paved road leaves the lot by an information kiosk. A sign here reads MUSCH TRAIL. Take this to begin with – it will quickly lead you to the Backbone Trail.

> **SIDE TRIP:** *While you're at Trippet Ranch, it's worth the walk of a few minutes to check out the visitors center, which is in the old skeet lodge left over from the days when this place was an actual ranch. The center is frequently closed; check the sign at the beginning of the walk for details.*
>
> *To reach the center, walk up the dirt road that leaves from the southeast corner of the parking lot[2] (the corner to the right of the one with the Backbone Trailhead). In less than a minute, bear left at the junction[3] instead of following the main road to the right. You'll reach the visitors center[4] within another minute.*

**The hike:** After leaving the MUSCH TRAIL sign, about a minute into the walk you'll cross a bridge over a seasonal creek by an old pond, then immediately come to a junction[5] with the Backbone Trail. (No sign indicates it's the Backbone Trail; straight ahead reads MUSCH TRAIL, and to the left is signed DEAD HORSE TRAIL. This signing issue stays consistent throughout the hike. There are practically no markers along this segment that read "Backbone Trail." Any sign that reads "Dead Horse Trail" is synonymous with the Backbone Trail.)

Turn left onto the Backbone/Dead Horse Trail. The footpath (named for a horse carcass discovered during the trail-building process) follows an old split-rail fence as it skirts the edge of a large meadow, occasionally winding in and out of a beautiful oak canopy on the left. This meadow is one of the best places on the entire Backbone Trail for spotting deer and other wildlife – squirrels, rabbits, chipmunks, coyote, and other critters. During our last visit, we found a doe and her fawns lounging around amidst the trees just off the trail.

At .3 miles, you'll come to a junction[6] at the end of the fence. The right fork (closed) leads across the meadow; instead continue straight and downhill on the Dead Horse Trail/Backbone Trail. The vegetation shifts from meadows and oaks to chaparral, offering less shade and somewhat more run-of-the-mill scenery. You're entering the sunniest stretch of the hike where, for the next ⅓ mile, the trail crosses a plateau carpeted with chaparral, chiefly manzanita and the spindly, evergreen-like chamise. Underfoot is mix of sandstone and semi-hardened sand, creating a nice crunch with nearly every step. Be wary of several steps as you start down this section.

In about two minutes, you'll come to another fork[7] at .4 miles. A sign on your right says DEAD HORSE TRAIL TO DEAD HORSE PARKING LOT, while one on the left reads 92 TRAIL TO ENTRADA ROAD .2 MILES. Turning left here would take you back to the parking lot via the 92 Trail (see alternate route below). Follow the Backbone Trail to the right.

*ALTERNATE ROUTE: If, at this point, you are returning from an out-and-back hike and would like a shadier alternate route back to the Trippet Ranch Parking Lot, you can veer off the Backbone Trail here and return via a lovely path with the odd name of "The 92 Trail." Instead of the large rolling meadow along the Dead Horse Trail, you'll enjoy more deep oak woodlands and some smaller meadows on this route. You're more likely to see deer on this short segment.*

*To follow this route, turn right (assuming you're heading back towards Trippet Ranch) onto the 92 Trail. The footpath, named for the 92-acre tract of land which it crosses, soon leaves the chaparral, traverses a meadow, then heads through oak woods along an old dam. After crossing a bridge over a disused spillway, the trail comes to a junction[8] .15 miles into the alternate route. The 92 Trail heads straight towards Entrada Road, while a smaller footpath heads off to the left. Take the left path.*

*You're now on "The 92 Spur Trail" (not to be confused with "The 92 Trail." Who names these things?). Follow the trail through deeper oak woods, then across a second meadow – a great spot for deer watching – then back into woods again. As you cross a wooded ridge between two depressions at about .3 miles, look for the rusting remains of an old floodgate on the right. The basin here was once a reservoir and the metal floodgate was used to lift a massive "drain plug" (which is still visible down below) to cover the pipe under the ridge.*

*Follow the trail uphill across another small meadow toward the park entrance booth. At .4 miles, the trail terminates at the park road directly across from the booth.[9] A sign here reads 92 SPUR TRAIL TO 92 TRAIL .3 MILES. To return to the Trippet Ranch Parking Lot, turn left and walk down the road another 1/10 mile.*

After leaving the 92 Trail junction, the Backbone Trail continues to meander across the sandy plateau, offering excellent views looking across Topanga Canyon to the much higher lands of Fossil Ridge (with its tower atop) and dramatic Hondo Canyon, where the Backbone Trail heads in a later segment.

At about ½ mile, beware of accidentally bearing right onto a side path[10] that heads down to join Robinson Road. The Backbone Trail goes to the left and is signed DEAD HORSE TRAIL TO DEAD HORSE PARKING LOT at this point.

*Down the steps and through the oaks above the footbridge*

Shortly after, the trail leaves the little plateau and for the next quarter mile generally follows the ravine of a seasonal stream as it heads down toward Trippet Creek. You'll probably appreciate the intermittent shade as you enter moderately wooded patches of oak and toyon on your steepening descent. You'll hit a series of railroad-tie steps, and in a few minutes, you'll be able to see some views through the branches into the canyon on your right. You may also begin to hear the sounds of Topanga Canyon Boulevard ahead.

.8 miles into the hike, you'll descend another set of intermittent steps, some of which are moderately steep. After a few minutes of steps, the forest closes in tighter as you pass under several large old oaks, their massive branches requiring a bit of ducking.

Soon the trail reaches the canyon bottom and crosses a sturdy wooden equestrian-and-hiker footbridge above Trippet Creek, which trickles in a sycamore and bay shaded glen. This may be the loveliest spot on the hike: water collects in tiny pools around the boulder-strewn creek, the canyon slope is covered with ferns, frogs can often be heard croaking below, and on a hot day the bridge is the perfect spot to experience nature's air conditioning from a cool breeze that often cascades down the gorge.

After crossing the bridge, you'll begin a brief climb that includes several steps. You'll reach the top of the climb at a small divide (.9 miles), but here the Backbone Trail gets tricky when it meets a smaller trail, the Summit Trail, and momentarily joins up with it. To stay on the Backbone Trail, turn left at the first junction[11] atop the ridge (the Summit Trail to the right heads down to a housing development). A few steps later, at the second junction,[12] turn right and follow the Backbone Trail as it heads down more steps into the forest. (Following the left fork here along the ridgetop would take you southbound on the Summit Trail to Entrada Road.)

Continue descending through a mixed forest offering intermittent shade until mile one, where a small side path[13] briefly joins the Backbone Trail as it crosses a seasonal stream on a large old rusted pipe. Continue straight on the main route, avoiding the side path where it joins up from the right and, a few steps later, veers off to the left.

About a minute later, you may notice a house that at first seems to be directly adjacent to the trail, as if you could practically walk right up to it, but in actuality it's part of a development perched on the edge of the mountain across a ravine. This spot is about the closest that the entire Backbone Trail comes to anybody's home.

You'll reach another junction[14] at 1.05 miles, where the trail exits the woods into an open area of waist-high scrub. The spur path splitting off to the right heads down to join a driveway at the end of Glen Trail. The Backbone Trail changes character here, widening into an old road and becoming rocky and steep.

The rocks underfoot switch from the tan sandstone typifying the majority of the Santa Monica Mountains to basalt, a much harder black volcanic rock that is unusual for this area. The final descent from this point lacks virtually any shade, so prepare for a good bombarding of sun on hot days.

Within a minute or two, you will see the Dead Horse Parking Lot down the hillside to your

*The footbridge under the sycamores*

right. As you approach the lot, the trail forks.[15] Head to the right and down the path to the parking lot, avoiding the left fork that leads under some power lines.

At 1.15 miles, you will arrive at the Dead Horse Parking Lot,[16] which tends to be frequently closed. It's a paved lot with restroom facilities (also frequently closed) and a drinking fountain. This is a good spot for a picnic lunch at the table under a huge shady oak.

From the picnic area, you may wish to continue hiking the short distance to Topanga Canyon Boulevard to finish this segment or to continue on to the next leg. To do this, walk to the back of the parking lot and pick up the faint Backbone Trail beyond the restrooms. The pathway, which is slight but acceptably well trodden, takes you downhill, winding through a grove of trees, past brilliant green grass, moss, and some ferns and clover. It's a picturesque if short section, and at times there are plenty of butterflies and moths zipping amongst the bushes. At a little junction[17] in the forest, bear left, avoiding the smaller path to the right that heads north.

1.2 miles from where you started, the modest trail emerges onto Topanga Canyon Boulevard[18] almost directly across from Greenleaf Canyon Road. The segment ends here, a good turnaround point. To continue on to the next segment, cross busy Topanga Canyon Boulevard here, head to the left for a few steps, then turn right up Greenleaf Canyon Road.

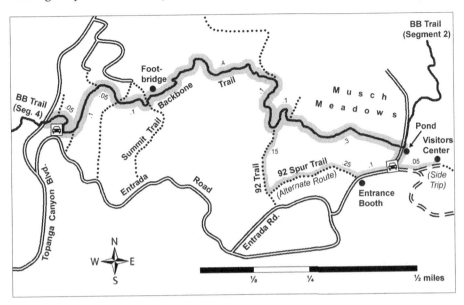

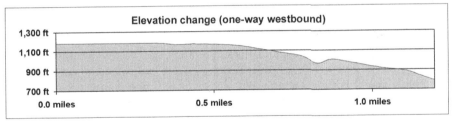

GPS COORDINATES

| | | |
|---|---|---|
| 1. N34° 5.613', W118° 35.249' | 2. N34° 5.595', W118° 35.240' | 3. N34° 5.590', W118° 35.203' |
| 4. N34° 5.599', W118° 35.160' | 5. N34° 5.652', W118° 35.229' | 6. N34° 5.693', W118° 35.513' |
| 7. N34° 5.675', W118° 35.546' | 8. N34° 5.544', W118° 35.560' | 9. N34° 5.580', W118° 35.357' |
| 10. N34° 5.771', W118° 35.546' | 11. N34° 5.734, W118° 35.846' | 12. N34° 5.727', W118° 35.848' |
| 13. N34° 5.728, W118° 35.907' | 14. N34° 5.750', W118° 35.982' | 15. N34° 5.684', W118° 36.022' |
| 16. N34° 5.678', W118° 36.050' | 17. N34° 5.701', W118° 36.060' | 18. N34° 5.694', W118° 36.091' |

## Segment 4:

# Topanga Canyon to Old Topanga Canyon
## ("Henry Ridge Crossing")

BBT segment length: .8 miles
Day hike length: .8 miles one-way
Suggested day hike: east to west, out-and-back

Elevation gain westbound: 218 feet
Elevation gain eastbound: 217 feet
Difficulty: easy to moderate – Shade factor: 70%

Bikes not allowed – Horses allowed – Dogs not allowed

East access: N34° 5.690', W118° 36.096'      West access: N34° 5.584', W118° 36.551'

This is the shortest segment of the Backbone Trail, but we include it as its own chapter because it is probably the most confusing segment as well. This brief series of connected horse trails, use paths, nature trails and dirt roads traversing Henry Ridge, the low divide between Topanga Canyon and Old Topanga Canyon, seems to have been thrown together as an afterthought. If you don't know the correct route, you may end up wandering onto the campus of Topanga Elementary School, heading the wrong direction up the ridgeline, or veering down a side road that will dump you out onto busy Topanga Canyon Boulevard.

Since the trail connects two major thoroughfares and passes just above a school, hikers looking for a wilderness experience won't find it here. On the plus side, most of this short hike is through beautiful woodlands with plenty of old oaks to shade your way, and part of the route is on an enjoyable nature trail with informative signs about indigenous flora. If nothing else, you'll find this segment fulfilling to hike if you are a fanatical Backbone Trail completist like we are.

This segment is short enough that many hikers simply add it onto a trip on adjoining Segments 3 or 5. Despite its brevity, the hike gets an easy-to-moderate rating because it's pretty steep in spots. There are no precipices to cross or rushing streams to hop – in fact, the most dangerous part of the hike is literally crossing Topanga Canyon Boulevard.

**Directions to trailhead:** From Pacific Coast Highway, drive north on Topanga Canyon Boulevard 4.8 miles. Just after passing Entrada Road on your right, park on the wide shoulder turnout along the northbound side of Topanga Canyon Boulevard, directly across the boulevard from Greenleaf Canyon Road at mile marker 4.80.[1] (If you're coming from the 101 freeway, exit Topanga Canyon Boulevard, then take it south for 7.9 miles and park in the same turnout.)

The turnout is large enough for several spaces of parallel parking, but if it's full there's also a small turnout parking area for two or three cars on Greenleaf Canyon Road just across the little bridge to the west of Topanga Canyon Boulevard. If all spaces are taken, you can always try parking at the Dead Horse Parking Lot about a minute's drive up Entrada Road. (The lot, which is frequently closed, also has a drinking fountain, a picnic table and flush toilets.)

**The hike:** Cross busy Topanga Canyon Boulevard and almost immediately turn right[2] (east) on sleepy little Greenleaf Canyon Road. As you cross the bridge, you will see Topanga Stream trickling down below. In less than a minute (.05 miles), turn left onto the signed Backbone Trail, which leaves just before a sharp right in the road.[3]

The trail heads up steeply from the get-go. This stretch also doubles as a nature trail maintained by a group called Keepers Of The Earth, and you might notice hand-made signs along it describing the local flora of the Topanga region.

*Coast live oaks along the trail*

The trail climbs the ridge steeply under the shade of twisted old oaks, aided by intermittent railroad-tie steps, as the traffic sounds from Topanga Canyon Boulevard slowly diminish with each step.

At about ¼ mile, after some steep climbing, you'll come to a fork[4] in the trail. A little sign saying BACKBONE TRAIL fails to inform you which way to go. If you were to head left, you'd be on a different nature trail that leads onto the property of Topanga Elementary School. Make a sharp right instead and head steeply uphill, climbing more railroad-tie steps.

Pass wooden stakes marked 5, 6, 7, and 8, continuing the steep ascent via steps, and finally leaving most of the sound of the boulevard behind. Soon you'll exit the oaks to climb alongside a steeply sloping grassy meadow. The trail heads up more steps as it skirts the edge of the meadow, with excellent views opening up from behind looking out across Upper Topanga Canyon. Eagle Springs Fire Road is visible heading up the ridgetop from Trippet Ranch on the other side of the canyon.

About a minute later you'll exit the field and reach the top of the divide between the two canyons, a small flat spot under beautiful old oaks. It's a great place to take a breather, grab a snack, and enjoy being out in the country.

At the top, the trail curves to the right. Follow it for less than a minute and you'll reach a T-shaped junction at .3 miles, where the Backbone Trail turns right.[5]

If you're using Google Maps' plotting of the Backbone Trail, be aware that it erroneously shows the BBT turning left here. (Although the route to the left is not technically the Backbone Trail, it does connect to the BBT in a few minutes – see "Alternate Route" below.)

> **ALTERNATE ROUTE:** *For a more complicated but woodsier and more pleasant route, you could take this smaller path, which will rejoin the Backbone Trail in a few minutes. Take the path to the left and head downhill through oak woods. In about a minute you'll continue through a sharp right switchback, and then cross above the amphitheater used by the elementary school (which will be down below to your left). Immediately after, head down some steps.*
>
> *At the bottom of the steps, you'll come to another "T"-shaped junction.[6] To the left leads to the school amphitheater – make a right instead. Within another minute, the winding trail dumps you out onto the Backbone Trail (aka Henry Ridge Motorway).[7] Turn left here to continue on the hike.*
>
> *(A note if you're using Google Maps: at this point, Google Maps erroneously shows the trail as crossing Henry Ridge Motorway and continuing directly down the hill to the water tanks below. In actuality, there's no path going that way and you might end up with a sprained ankle and several scratches if you try to bushwhack your way down it. Turn left instead.)*

From the T-junction just past the ridgetop, take the Backbone Trail to the right. Traverse the sun-exposed ridge for a minute and you'll come to an unmarked "Y"-shaped junction.[8] Take the left fork that goes slightly downhill; avoid the right fork that heads higher up the ridge. In about a half minute the trail joins dirt Henry Ridge Motorway[9] at .35 miles into the hike. Turn left, following the road downhill.

After a steep and sun-exposed descent of about two minutes, the dirt road ends by a chain gate at ½ mile.[10] The road becomes paved and it's obvious that you'd be entering the school property if you continued straight. Instead, take a sharp right onto a paved road heading towards some large water tanks.

About a minute down the road, the pavement ends. Follow the footpath that skirts just to the right of the tanks. The ugly tanks and adjacent razor wire fence make this short stretch one of the least picturesque of any along the entire Backbone Trail.

After leaving the tanks, the trail returns to more respectable scenery and begins angling its way down an old horse path. It descends the eastern side of Old Topanga Canyon through another oak grove, with the sound of Old Topanga Canyon Road growing louder as you continue. This section of trail is moderately steep and a bit rocky, so hiking boots are recommended.

Soon you'll exit the grove and descend more gently through a mix of shady oaks and sunny chaparral, getting some fine views across the canyon. Hondo Canyon, containing the next segment of the Backbone Trail, is visible ahead while Fossil Ridge (with tower atop) looms in the distance.

As you near Old Topanga Canyon Road, the trail veers to the left. Another minute of descent will get you to the guardrail along the road. Follow the rail until it opens onto the road next to a yellow Horse Crossing sign, at .8 miles.[11] Cross the road directly; using the small route which appears to cross under the road via a drainage underpass is not a good idea. (Don't ask how we know. Just trust us, we know.)

From here you can continue on to Segment 5 of the Backbone Trail directly across the road, or return the way you came.

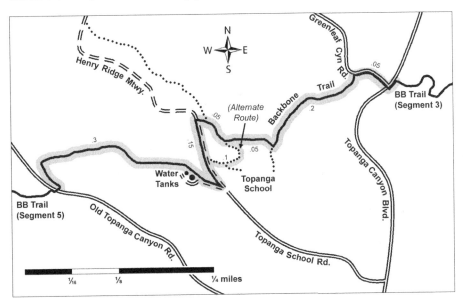

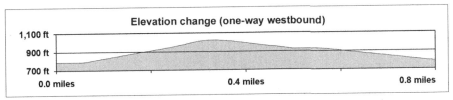

GPS COORDINATES

1. N34° 5.690', W118° 36.096'  2. N34° 5.685', W118° 36.108'  3. N34° 5.700', W118° 36.145'
4. N34° 5.629', W118° 36.251'  5. N34° 5.625', W118° 36.321'  6. N34° 5.603', W118° 36.324'
7. N34° 5.608', W118° 36.346'  8. N34° 5.644', W118° 36.341'  9. N34° 5.648', W118° 36.355'
10. N34° 5.576', W118° 36.324'  11. N34° 5.584', W118° 36.551'

## SEGMENT 5:

# OLD TOPANGA CANYON TO SADDLE PEAK ROAD
## ("HONDO CANYON TRAIL")

**BBT segment length: 3.7 miles**
**Day hike length: 3.7 miles one-way**
**Suggested day hike: east to west, out-and-back**

**Elevation gain westbound: 1,647 feet**
**Elevation gain eastbound: 75 feet**
**Difficulty: moderate – Shade factor: 70%**

**Bikes not allowed – Horses allowed – Dogs not allowed**

**East access: N34° 5.582', W118° 36.556'       West access: N34° 5.024', W118° 38.157'**

Deep woodlands, huge boulders, moss and ferns, spectacular cliffs, a babbling brook, dramatic views from lofty heights… this hike boasts an embarrassment of riches. The only challenge is that you'll pay for it with a hefty climb. With an elevation gain of over 1,600 feet, this stretch up Hondo Canyon comprises one of the biggest climbs on the Backbone Trail. Just *looking* down the canyon from the top of the trail at Saddle Peak Road can give you the dizzies. But it's worth every step. Walking the heights of the upper canyon rewards you with a real alpine feeling – a sense of being high in the mountains – all within a half hour's drive of Santa Monica.

This place is well named. Hondo (please don't call it Honda Canyon – it's not named after your car) is Spanish for "deep," and this is one of the most rugged and dramatic canyons in all of the Santa Monica Mountains. Ringed by a series of impressive cliffs, the upper part of the canyon affords excellent views of the Santa Suzanna Mountains, the San Fernando Valley, Topanga Canyon, and even a small peek at the Pacific Ocean about four miles away.

Along the way, the trail crosses rolling grasslands, ambles beneath canopies of oak and bay trees, climbs in and out of elfin forests amongst red-barked manzanitas,

*500-foot high cliffs across the upper canyon*

and winds around moss-covered boulders from a series of recent landslides that disrupted the construction of the trail during the Northridge Earthquake in 1994.

You'll pass old cabin ruins, cross streams, peer down into a purple gorge, and even pass the former site of a hidden pot farm… Hey, welcome to Topanga!

Despite its proximity to civilization, most of this hike feels pristine and wild. Plenty of deep, shady oak and bay woods, a wide variety of vegetation and interesting geology make this one of the most satisfying hikes on the entire BBT.

Although we prefer doing this segment as a half-day out-and-back hike, it could be shortened by using either a car shuttle, a second car or a rideshare service (see Chapter 3). Hiking back down, however, is easy and enjoyable.

**Directions to trailhead:** From Pacific Coast Highway, turn north onto Topanga Canyon Boulevard. After passing through Topanga village, turn left at the "Y"-shaped junction onto Old Topanga Canyon Road at 4.3 miles. Continue for another .4 miles. Just past the small bridge over the creek, park in the dirt turnout that holds about eight cars south of mile marker 5.94.

If you're coming from the 101 freeway, exit at Topanga Canyon Boulevard and take it south for 8.1 miles. Turn right onto Old Topanga Canyon Road and continue for another .4 miles. Just past mile marker 5.94, park in the dirt turnout that holds about eight cars. If full, a larger turnout is available north of mile marker 5.77.

**The hike:** Locate the signed trailhead[1] at the southern end of the turnout. The trail promptly heads downhill into the woods and within a minute, you'll come to Old Topanga Canyon Stream. Several routes are available to cross it on stones.

Immediately after the crossing, the trail heads up a set of steps past a sign informing you that you've entered Topanga State Park.

About .2 miles into the hike, after a couple of minutes crossing open grassland, you will find yourself back along a smaller stream in a shady riparian woodland. When the water is reasonably high, it helps produce a soothing, trickling effect.

Within a couple of minutes, you'll cross the stream over an old concrete ford, partially collapsed, with dilapidated pipes and bits of

*Rolling meadows and oak woods make up the first part of the hike*

rusted junk from days of "progress" gone by. Leave the stream, climbing up some wooden steps and through a large grassy field on a hill.

After about a minute of climbing up from the stream, at .3 miles the Backbone Trail makes a sharp left at a junction.[2] Don't continue straight on the old horse path.

A few minutes later (another 1/10 mile), angle left again at a second junction[3] by a metal sign with an arrow pointing left, avoiding a sharp right onto a smaller and lightly used path.

The trail continues climbing moderately up the north slope of the little canyon, which incidentally is not Hondo; you're in a smaller tributary of Topanga Canyon. The route works its way up sun-exposed hillsides until ½ mile, when it re-enters the forest of oaks, bays, and tall toyons. You'll pass an impressive rock formation which towers over the trail, then continue in and out of woodsy patches for the next several minutes, with the little stream getting steadily closer on your left.

At about .6 miles, a smaller path[4] leads sharply down to the left, crossing the stream and heading up into a housing community. Instead, continue uphill on the Backbone Trail.

For the next ¼ mile, the trail parallels the streambed up the small canyon, skirting grass-lands and oak woods, growing progressively steeper as it ascends. Eventually it becomes downright steep, then

*Looking up from the bottom of the purple gorge*

turns northward, finally leaving the woodsy stream for good as it climbs out of the canyon up a short thigh buster – probably the steepest stint on the entire hike.

Then, at .8 miles, you'll top a small, grass-covered, windswept divide between the two canyons, where you'll reach a minor "T"-shaped junction.[5] Take a left along the ridgeline, avoiding the smaller trail that heads right. The ridgetop offers your first views down into Hondo Canyon on the right.

About a minute later, you'll begin descending down into Hondo Canyon – the longest downhill stretch of the hike. The overhanging bluffs here support a virtual garden of elfin forest vegetation: hollyleaf cherry, western mountain mahogany, large toyon, laurel sumac and ceanothus, plus some occasional ferns and moss are among the wide variety of plants thriving here. Ahead you'll get terrific (or is it terrifying?) views of Hondo Canyon looming up in front of you, with its dramatic cliffs near the top. Yep, you're heading up there!

You will end your descent at 1.1 miles, leaving behind the last views of farm houses and civilization as you approach a little narrows of pretty purple cliffs and rock formations. The distinct purple hue of the trail here comes from the Sespe Formation sandstone deposited by freshwater seas 30 to 40 million years ago.

**SIDE TRIP:** At 1.2 miles a very short side trail spurs off to the right,[6] heading to an overlook rock which hangs above the stream and the purple gorge, offering fine views both up and down Hondo Canyon. It's definitely worth the half-minute detour. The chamise-speckled rock is mostly sun-exposed but makes a good picnic spot nonetheless.

*SIDE TRIP: From the overlook, a short trip of a few minutes down a somewhat precarious side path will take you to the bottom of the narrows, with its beautiful stream-scoured purple walls and babbling brook.*

*To get there, take the light use path that leaves just north of the overlook rock. Exercise caution on the steep and gravelly sections, especially when near drop-offs. If the path looks questionable or if you are unsure-footed, please don't attempt it.*

*In about a minute you'll cross a side tributary in a little purple flume. Though usually dry, enough water occasionally comes down here to scour the rocks into smooth shapes and to support moss. The flume's alternating layers of purple sedimentary rock and conglomerate, multicolored stones resemble the layers of a cake and provide a crash course in the local geology.*

*The path passes below a 10-foot cascade (usually dry), heads down the flume for a few seconds, then turns left into the woods and terminates in the narrows at the bottom of Hondo Canyon in another half minute. The narrows are a similar, if larger, version of the flume. Beautifully smooth purple slopes rise up on both sides, with potholes scoured by whirlpools long gone. The overlook rock towers above.*

*Return the way you came, taking care on the steep and slippery path.*

After the lookout rock, you will begin climbing Hondo Canyon in earnest. The trail presses fairly steeply uphill through a chaparral mix of chamise, ceanothus, manzanita, yucca, sage and grass. This is the most arid stretch of the hike, and it can be quite sun-exposed at times, particularly while passing through a patch of shoulder-high manzanita and chamise at about 1⅓ miles. At other times, taller patches of elfin forest can cast some shade.

The trail climbs the south slope of the canyon, leaving the stream far below. Little views begin to open up looking back down the canyon toward Topanga.

Eventually the trail enters thick oak and bay woods (about 1½ miles into the hike). These woods will last for the next two miles, interrupted only by the intermittant patch of chaparral.

About a minute after entering the woods, the trail passes through a rusting gate along a wire fence, traversing a few side ravines in an area shaded by bay trees with planted eucalyptus and delicate ferns. The ruins here (stone foundations mostly) are all that is left of the cabin that belonged to nutritionist and wellness movement pioneer Paul Bragg. The cabin burned, along with much of Malibu, in the infamous Old Topanga Fire of 1993.

In another minute you'll pass a tiny path[7] spurring off to the right at 1.6 miles. (The path heads for about a minute along a little marshy tableland, fed by a spring that sustains some cute patches of clover. It ends at an unremarkable lookout over the canyon and may require a push through poison oak, so it isn't recommended.)

From here, the Backbone Trail begins the first of two major sets of switchbacks. This first set will take you up the canyon at a moderate incline for about ⅓ mile, under a shady canopy of oaks and bays, past plenty of ferns, grass, and poison oak.

Most of the way up the switchbacks (1.8 miles), you'll pass a landmark on the left: a large rock that appears to be split. A couple of good sitting rocks on the trail next to it provide a rest opportunity.

By the two-mile mark, the trail has finished its first major set of switchbacks and straightens out for a gentle climb. Clinging to the ridge halfway up the canyon, it offers spectacular views looking to the right and back across the Topanga Basin. With good visibility you can even see the Santa Suzanna and San Gabriel mountain ranges to the north of the San Fernando Valley. You'll also get views of the old Topanga Lookout site, appearing as a promontory jutting up ahead at the top of Hondo Canyon.

The trail climbs briefly up a side canyon at 2.2 miles, offering more impressive views of the cliffs that overhang the top of Hondo. Shortly thereafter, you'll make your first of a few crossings through the remnants of a landslide that came down in the 1994 Northridge Earthquake, which temporarily hampered construction of the trail. Tumbled boulders and the remnants of rotting old crushed oaks are all that remain from the event. With every passing year, nature reclaims a bit more of the remnants of the landslide, making it seem less catastrophic.

*A deep bay tree forest along the upper switchbacks*

At 2.4 miles, you will again find yourself close to Hondo Stream, which often babbles in a beautiful glen directly below on your right amidst a particularly deep bay tree forest... only to abruptly leave the stream behind as the trail embarks on its second major series of switchbacks up the canyon's southern escarpment.

After several minutes of ascending these reasonably moderate switchbacks through pleasant woods, as the trail turns back up-canyon you'll begin to see the humungous abandoned microwave communications tower beyond the canyon on top of Fossil Ridge.

You'll enjoy more terrific views looking down the canyon, and from these higher viewpoints it becomes apparent just how rugged this place is. Across the upper canyon are its 500-foot high cliffs, tan and golden at the top, with a purple promontory below.

Higher up the switchbacks (2.8 miles), look for a welcome surprise: a little moss-covered grotto. It's actually a seasonal streambed kept perennially moist under constant shade from big beautiful oaks. A minute later, after another switch-back, you'll get an encore as you cross the little grotto again. Here you will finally leave the second major series of switchbacks.

A few more minutes takes you to a rare and short downhill stretch through chaparral, with more excellent views of the colorful cliffs on the opposite side of the canyon.

You'll enter a third and smaller series of switchbacks near the top of the canyon at mile three. The forest here gets even deeper and lusher, with old oak and bay trees, moss-covered boulders, ferns and poison oak everywhere in a cool and moist setting.

After about five minutes you'll cross three small streams in rapid succession – it was in this general vicinity that the aforementioned pot farm was discovered a few decades ago. A minute or two later, look for an interesting anvil-shaped rock formation that towers over the trail on the right. It's like a gate that marks the entrance to a particularly boulder-strewn patch where the trail becomes very rocky, steep and slow-going. Every now and then, occasional steps made from placed rocks aid in the climb.

At 3.4 miles you'll cross another stream by a pair of large pink boulders. It might seem as if you've finally switched to the north side of the canyon, but not so fast. The main stream still lies ahead, after about two minutes of walking through jungle-thick elfin forest which has been meticulously cut back by thoughtful trail stewards.

Within another couple of minutes (3½ miles), you'll cross Hondo Stream for good and switch to the north side of the canyon. Like most south-facing slopes, it's more heavily bombarded by the sun, and as you cross the stream the vegetation immediately changes to an arid mix of grass and low chaparral. This lower growth opens up spectacular views looking down the canyon. From the vistas along this stretch, you can see vast parts of the San Fernando Valley in the distance and just a bit of the Pacific Ocean as well. Almost all of Topanga Canyon is spread out before you.

A sharp bend to the left at 3.6 miles reveals the same imposing cliffs that have loomed ahead for most of the hike, only now you are finally skirting the base of them, offering a wonderful up-close perspective on the palisade. Shortly after, a small use path[8] used by climbers heads off to the right and up towards the cliffs, but quickly degrades into a very rough bushwhack.

Continue moderately uphill on the final stretch of trail, in and out of the shade of tall chaparral, with frequent views looking back down the majority of the canyon.

You'll reach a shade-covered junction[9] with the Fossil Ridge Trail (the next segment of the Backbone Trail) at 3.7 miles. From here you can continue to the right on to Segment 6, or return the way you came to complete an out-and-back hike of 7.4 miles.

Just a few seconds of walking down the side path to the left will lead you to the trail's terminus at Saddle Peak Road,[10] where there is very limited parking. The views looking back down Hondo Canyon from the road should impress you as to just how far you have climbed.

*SIDE TRIP: If you're looking for some ocean views or a spot to take a celebratory breather, there's a decent picnic rock another four minutes further up the Backbone Trail (see Segment 6 description). The large, flat boulder offers rewarding views of the Pacific.*

***SIDE TRIP:*** *Hardier hikers often continue along the short upcoming Segment 6, and then on to the Topanga Lookout site. From the lookout, you can gaze down on all of Hondo Canyon as well as most of the Los Angeles area, making for a very rewarding end to a fairly challenging hike. To get there, continue on through the next segment until the trail ends, then turn right on the paved road and follow it to the lookout (see side trip under Segment 6).*

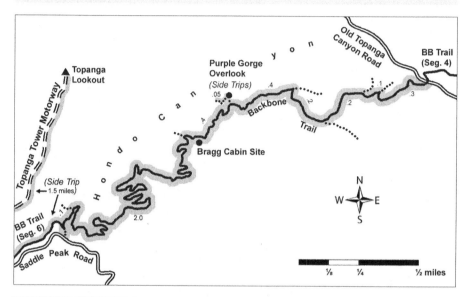

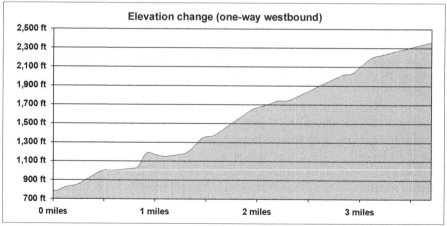

GPS COORDINATES

1. N34° 5.582', W118° 36.556'    2. N34° 5.528', W118° 36.768'    3. N34° 5.536', W118° 36.840'
4. N34° 5.440', W118° 37.019'    5. N34° 5.552', W118° 37.143'    6. N34° 5.500', W118° 37.417'
7. N34° 5.373', W118° 37.608'    8. N34° 5.119', W118° 38.111'    9. N34° 5.032', W118° 38.155'
10. N34° 5.024', W118° 38.157'

## SEGMENT 6:
# SADDLE PEAK ROAD TO LOIS EWEN OVERLOOK
### ("FOSSIL RIDGE TRAIL")

**BBT segment length: .7 miles**
**Day hike length: .7 miles one-way**
**Suggested day hike: west to east, out-and-back**

**Elevation gain westbound: 133 feet**
**Elevation gain eastbound: 118 feet**
**Difficulty: easy – Shade factor: 80%**

**Bikes not allowed – Horses allowed – Dogs not allowed**

**East access: N34° 5.024', W118° 38.157'     West access: N34° 4.878', W118° 38.734'**

This brief and easy segment, also known as the Fossil Ridge Trail, delivers plenty of bang for your hiking buck. Why? Because it's mostly level easy strolling, but since you start at an elevation over 2,300 feet, you'll feel like you've hiked up a mountain without having to actually hike up anything. It's particularly good for kids and people who don't want to climb very much.

The short trail heads through a miniature forest of tall overhanging chaparral, mostly ceanothus and laurel sumac, offering plenty of cool shade, chirping birds, perpetual greenery and interesting rock formations. Every once in a while you can peek through the forest overhang and see incredible views far out over the Pacific Ocean, which is over a half a mile below, as well as Catalina Island and the Palos Verdes Peninsula. You may also spot fossils (for which the ridge is named) dating back some 30 to 40 million years embedded in the pink Sespe Formation sandstone. Several of these fossils have the obvious shape of shell indentations.

*Fossils along the trail*

Unfortunately, this hike comes with a sad caveat. In recent years the area has become something of a haven for graffiti taggers who have made a hobby of spray painting every possible square inch of the abandoned microwave communications tower on top of Fossil Ridge. Lately, some of the more misguided of these individuals have taken to tagging the rocks along the Fossil Ridge Trail, and the area now has an alarming amount of graffiti.

When it's on an abandoned structure, call it self-expression, call it vandalism, call it art... but when they start tagging the Backbone Trail, we call it nauseating. What's especially annoying about the graffiti is that most of it isn't even name-writing or territory marking from at-risk youth or gangbangers seeking a way to feel important in an indifferent world. The graffiti here is just stupid stuff like smiley faces, hearts and phalluses – it looks like it's been drawn by smug, well-off teenagers from the Westside and The Valley who are bored on a Friday night. Why the county hasn't dismantled the "Topanga Graffiti Tower" is a mystery to us, but if you do catch anyone tagging rocks along the Backbone Trail, you have our blessing to grab their paint and spray "get a life" on their forehead.

This segment is so short that many hikers choose to combine it with an adjacent segment or two. A particularly popular thigh burner is to start at Old Topanga Canyon Road and hike up Hondo Canyon (aka Segment 5), continue

*Half a mile above the Pacific with no climbing required*

westbound on the Fossil Ridge Trail, then turn right and head out to the Topanga Lookout (see side trip), from which you can look down into Hondo Canyon and admire your accomplishment.

**Directions to trailhead:** From Los Angeles, take Pacific Coast Highway west. Once you have passed Topanga Canyon Boulevard, continue west for another 3.4 miles, then turn right onto Las Flores Canyon Road (signed simply as LAS FLORES). Drive up the canyon for 3.5 miles, then turn right onto Rambla Pacifico. After another 0.7 miles, turn right again at Schueren Road. In another 1.8 miles, at the top of the pass where Schueren intersects Saddle Peak Road and Stunt Road, park in the paved parking lot for the Lois Ewen Overlook.[1]

If you're coming from the 101 freeway, exit Mulholland Drive in Calabasas and take it south for .6 miles. Turn right on Valmar Road (aka Old Topanga Canyon Road) and follow it 1.1 miles. Turn right on Mulholland Highway, go another 3.9 miles, then turn left on Stunt Road. Take Stunt up the mountain for another 4.1

miles. Then, at the top of the pass where Stunt intersects Saddle Peak and Schueren Roads, park in the paved parking lot for the Lois Ewen Overlook.[1] (If the lot is full, try the overflow parking area 1/10 mile down Stunt Road at mile marker 3.99.)

**The hike:** Start off by taking a minute to enjoy the views from the 2,375-foot elevation of the Lois Ewen Overlook, named for a mayor and environmentally-conscious public servant who helped establish the Mountains Restoration Trust. It can get pretty crowded on weekends with folks coming and going.

*One of Fossil Ridge's miniature tree tunnels*

From the overlook parking lot, head east on Saddle Peak Road and almost immediately veer left onto a smaller and rarely used paved road, Topanga Tower Motorway. Walk around the gate. (Please avoid the use path that shortcuts up the ridge. Such paths create erosion problems.) On a clear day, you should already be enjoying first-rate views of the distant peaks of the Los Padres National Forest north of Ventura and, farther to your left, the Santa Ynez Mountains above Santa Barbara, fifty miles away.

Head down the old road for about a minute and a half (about 1/10 mile), then turn right onto the single-track Backbone Trail. If you're not careful, it can be easy to miss the sign at the trail entrance[2] that reads FOSSIL RIDGE TRAIL, HONDO CANYON TRAIL .6 MILES, next to one of the electrical poles lining the road.

*SIDE TRIP: Continuing straight on the paved road will take you to the Topanga Lookout, the former site of a fire lookout tower based on a high promontory which juts out over Hondo Canyon. From the concrete platform you'll get a 360° bird's eye panorama of most of Los Angeles, including ridge after ridge stretching for a good 50 miles in every direction. On a clear day the view extends 100 miles to Mount San Gorgonio. It's the kind of neck-stretching panorama that photos just can't do justice.*

*The 1.8-mile round-trip to the lookout is a pretty easy walk along a road – paved at first, then dirt – that undulates up and down along the ridge but has little real elevation change. For the entire side trip, you'll be walking through the Cold Creek Canyon Preserve owned by the Mountains Restoration Trust.*

*To reach the lookout, continue straight on the paved road from the BBT junction. The road surface is scribbled with observational graffiti, some of which is clever, but most of which is on the level of "Skeeter sucks" or "I'm so high." After a quarter mile of gentle uphill, the road reaches a "Y"-shaped junction. The paved road heads to the right up to the abandoned communications tower. Instead, take the dirt road to the left.*

*The road descends gently and the graffiti stops (no more pavement to tag). In another quarter mile, it crosses a saddle and climbs again. The area becomes so sandy that it nearly resembles dunes. Soon you'll head down along the north side of the ridge.*

*.9 miles from where you left the Backbone Trail, you'll arrive just below the base of the old lookout tower platform. (Although the road continues, this should be your destination.) To access the lookout, either scramble directly up a short but very steep path to the top, or take the small trail to your right that leads there at a much gentler pace, although its sharp drop-off over Hondo Canyon might give you the willies.*

*There's a ton of graffiti on the concrete platform, as well as other detritus left behind by an army of drunken melonheads. Ignore it or shake your head in bemusement, but don't let it detract from the stunning view.*

The Backbone Trail wastes no time taking you up a short but steep ascent onto the ocean view side of Fossil Ridge. As you climb the hill (and pass the aforementioned use path which comes in from the right[3]), you'll get expansive views looking down Las Flores Canyon and, on a clear day, over the Palos Verdes Peninsula, which offers a rare perspective revealing that Los Angeles is actually on a large bay. Looking back and to the north, you might be able to see the Topa Topa Mountains near Ojai and the high peaks of Ventura County near Frazier Park.

The trail flattens out after just a few minutes, then soon leads you into a little elfin forest of tall ceanothus, offering frequent opportunities to peek through the forest overhang and see some beautiful Malibu estates down below.

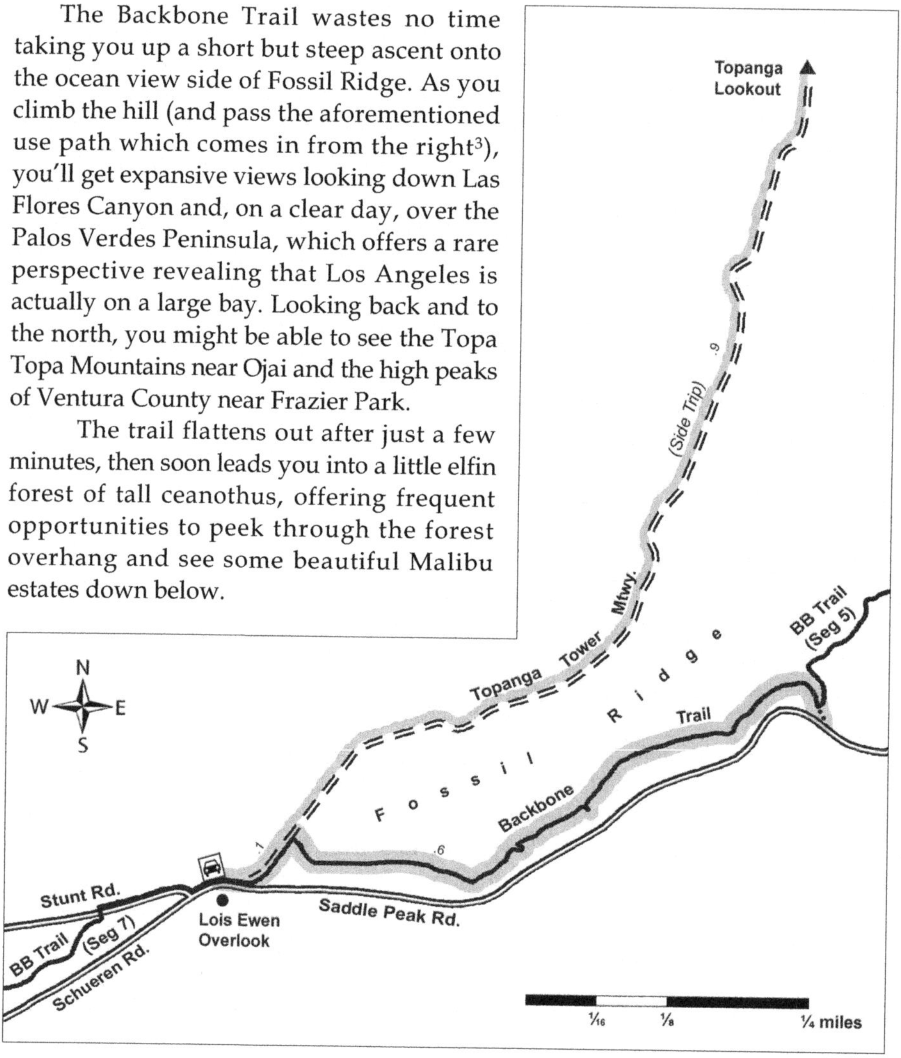
Topanga
Lookout
.9
(Side Trip)
Topanga Tower Mtwy.
BB Trail
(Seg 5)
Fossil Ridge
Backbone Trail
.6
.1
N
W
E
S
Stunt Rd.
Lois Ewen
Overlook
Saddle Peak Rd.
BB Trail
(Seg 7)
Schueren Rd.
1/16
1/8
1/4 miles

At ¼ mile, as you work your way through the miniature forest, you will come to a single switchback. A few minutes later (.4 miles), the trail begins its descent with another small switchback.

Soon you'll come across a large old pipe leading down the mountain, carrying cables to the radio transmitter tower on the peak above. Fortunately, the tower is all but invisible from this trail, maintaining the impression of being in a semi-wilderness.

At .6 miles, you'll come across a large rock outcrop that makes a good rest spot. From here you can see the front range of the San Gabriel Mountains over 30 miles away. You'll also get a good view of Hondo Canyon, the next segment of the Backbone Trail as it heads east.

As you descend past a break between the hills on the right, you'll be able to look far out over the ocean, with views of Catalina Island and, on a clear day, pointy little Santa Barbara Island to the right of Catalina.

This short hike ends at a fork in the trail[4] just .7 miles from where you started. If you head left here, you'll continue downhill on Segment 5 of the Backbone Trail into Hondo Canyon. It's clearly marked with a sign that reads OLD TOPANGA CANYON ROAD 3.6 MILES, PART OF THE BACKBONE TRAIL SYSTEM. If you go right, you will almost immediately emerge from the woods onto Saddle Peak Road,[5] where there are some turnout parking spaces and a good view down into Hondo Canyon.

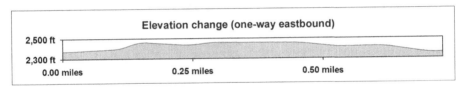

GPS COORDINATES
1. N34° 4.878', W118° 38.734'     2. N34° 4.917', W118° 38.647'     3. N34° 4.893', W118° 38.630'
4. N34° 5.032', W118° 38.155'     5. N34° 5.024', W118° 38.157'

# LOIS EWEN OVERLOOK TO SADDLE PEAK
## ("SADDLE PEAK TRAIL" EAST APPROACH)

**BBT segment length: .7 miles**
**Day hike length: .9 miles one-way**
**Suggested day hike: east to west, out-and-back**

**Elevation gain westbound (to summit): 498 feet**
**Elevation gain eastbound (from summit): 68 feet**
**Difficulty: easy to moderate – Shade factor: 60%**

**Bikes not allowed – Horses allowed – Dogs not allowed**

**Access: N34° 4.878', W118° 38.734'**

Saddle Peak, as the name would suggest, is a double peak with a high saddle in-between overlooking the Pacific Ocean. While the west summit is topped with a small city of communications towers, the east summit is undeveloped, windswept, and wonderful. From its 2,805-foot height you can see most of Los Angeles stretching out to the horizon.

*Backbone Trail along the cliffs just below the summit*

Your view includes Catalina Island, Mount Baldy, the San Gabriel high peaks, the Santa Ana Mountains down in Orange County, plus seemingly endless valleys, and of course the glorious ring of beaches along the Santa Monica Bay stretching twenty miles from Malibu to Palos Verdes. The flat top of Saddle Peak makes the perfect place to sit down and enjoy a picnic, with no cover charge for your multi-million dollar view.

This hike is the easy way to the summit. With a relatively brief climb of 500 feet over about a half hour, you can reach the top of one of the highest peaks in the Santa Monica Mountains. After an underwhelming start past an ugly water tank, you'll find yourself on a lovely alpine path ascending the perpetually shaded north side of the mountain beneath overhanging cliffs, through patches of woods with perennially green grass, past moss-covered rocks and even the occasional mushroom patch.

This segment is one of the oldest parts of the Backbone Trail, dating at least as far back as a 1939 Los Angeles Times hiking column which describes it in "that part of the Santa Monica Mountains known as the Malibu." Since it's a "destination" hike

*Unforgettable view from the summit*

to a peak, there's no car access at the end of the segment. From there, you'll need to turn around and hike back another .9 miles or follow through on to the next segment.

**Directions to trailhead:** From Los Angeles, take Pacific Coast Highway west. 3.4 miles past Topanga Canyon Boulevard, turn right onto Las Flores Canyon Road (signed simply as LAS FLORES). Drive up the canyon, turning right on Rambla Pacifico at 3.5 miles. In another 0.7 miles, turn right again at Schueren Road and follow it for 1.8 miles to the top of a pass where it intersects with Saddle Peak Road and Stunt Road. Park in the paved parking lot for the Lois Ewen Overlook.[1]

If you're coming from the 101 freeway, exit Mulholland Drive in Calabasas and take it south for .6 miles. Turn right on Valmar Road (aka Old Topanga Canyon Road) and take it 1.1 miles. Turn right on Mulholland Highway, go another 3.9 miles, then turn left on Stunt Road. Take Stunt up the mountain for 4.1 miles. At the top of the pass where Stunt intersects with Saddle Peak Road and Schueren Road, park in the paved parking lot for the Lois Ewen Overlook.[1] If the lot is full, try the overflow parking area $^1/_{10}$ mile down Stunt Road at mile marker 3.99.

**The hike:** Leave the overlook and walk west on Saddle Peak Road, immediately angling right at the fork onto Stunt Road. (Avoid the use path that scrambles up the ridge directly across from the parking lot, erroneously labeled on some maps as the BBT.) Walk downhill along the broad north shoulder of Stunt Road, and in about two minutes ($^1/_{10}$ mile) you should see the single-track Backbone Trail[2] leave from the opposite side of the road across from mile marker 3.99. A sign here reads SADDLE PEAK TRAIL. PIUMA RIDGE TRAIL 4.8 MILES, MESA PEAK MOTORWAY 6.7 MILES.

The trail begins climbing the ridge between Stunt and Schueren Roads. It tops the crest at .2 miles, allowing views both southward over the Santa Monica Bay and northward across the Conejo Valley.

Just before the crest, a use path[3] heads sharply to the left along the ridgeline in the direction of the parking lot. It's erroneously labeled on some maps as the Backbone Trail and on your return trip you might be tempted to use it as a shortcut, but please don't. It offers nothing special in the way of scenery and it eventually deteriorates, ending with a hazardously steep climb down a crumbly embankment.

As you continue on the Backbone Trail up the ridge, most of your ocean views are blocked by a lower ridge and chaparral... but plenty of views are soon to come.

After a few more minutes you should see a water tank looming ahead. Head toward the tank, climbing an open stretch along the ridge, which offers excellent views looking back over Santa Monica, much of Los Angeles, and the Santa Monica Bay.

At .4 miles, you'll rejoin the crest of the ridge, where you'll reach a junction[4] with Mildas Drive, an old paved road that heads up to

*Looking north from the top*

the water tank. A little marker on the left side of the road reads BACKBONE TRAIL. Follow the road directly up to the water tank, skirting around the tank's left side.

Once past the tank, continue up the old paved path as it becomes less well-maintained and moderately steep. As you climb away from the tank, the toyon, ceanothus and laurel sumac grow higher and close in tighter, creating more shade.

After about five to ten minutes, the vegetation transitions to a proper forest. The trail becomes overhung with oaks and the occasional bay tree, blocking your view but providing welcome shade on a hot day. From here onward, your hike will be much gentler, shadier, and quite pleasant. The trail (actually the far-gone remnants of an old road), clings to the north side of the mountain just below a range of sandstone cliffs that block the sun most of the time. The grass in this perpetually shaded stretch is usually green, the trail is often moist, the rocks are covered with moss, and the occasional fern grows. On a winter's day, hiking this stretch can feel almost – and this is rare for Southern California – gloomy. The route becomes quite rocky below the cliffs, necessitating hiking boots.

At .7 miles, just after you leave the shade of the cliffs, you'll reach a junction[5] at a local high point on the Backbone Trail of 2,700 feet. The BBT continues to the right, entering Malibu Creek State Park and descending the north slope of Saddle Peak (see next chapter). To reach the summit, take the spur trail that leads off to the left.

*SIDE TRIP: Continuing westward along the Backbone Trail for another .2 miles will take you down through an interesting area known as the Rock Garden. It's a green patch strewn with sandstone rock pinnacles, many of which have a pink tint to them reminiscent of Arizona or Utah, honeycombed with cavities.*

*To reach the Rock Garden, simply continue to the right on the Backbone Trail, following it as it heads steeply downhill through the rocks. When you exit the lower end of the garden, turn around and head back up. (For more detailed description, see side trip box in Segment 8.)*

From the junction, the spur trail takes you to the summit of Saddle Peak in about five minutes up a sun-exposed moderate incline of about 100 feet. Even if the summit is not your ultimate destination, the views from the top are well worth the detour.

A few minutes up the spur trail, as you cross the "saddle" atop Saddle Peak, you'll notice a bizarre lone hoodoo (a rock tower) standing right in the middle of the saddle and on the lip of the 2,700-foot drop towards the Pacific. From a distance it may resemble the ruins of an ancient watch tower.

At .8 miles, you'll reach another junction.[6] To the right will take you down to a lower overlook and on to the rock hoodoo.

*Lone Hoodoo at sunset*

*SIDE TRIP: To reach the Lower Overlook and lone hoodoo, take the right fork and walk downhill along the old road for about a minute. Here a short path heads to the left[7] and uphill to the overlook.*

*From the Lower Overlook you'll enjoy views of the ocean, the city, Palos Verdes and Catalina Island, and you'll get a different perspective looking more sharply down into the canyons just below Saddle Peak. To the west, you'll get a glimpse of waves breaking at Malibu Lagoon. A series of smooth boulders along the rim of the overlook (both to the left and the right) offer several good sitting spots. If you've been holding off on having a picnic lunch, it isn't going to get better than this.*

*After heading back down to the dirt road, if you continue farther along the road for about another minute, you'll get to the lone hoodoo, which is interesting enough but not a must-see. The road continues toward the small city of transmission towers atop the west summit, which is closed-off private property, precluding any views to the west.*

To access Saddle Peak's east summit, turn left at the junction and continue uphill. You'll reach the windswept top about .9 miles after you left the Lois Ewen Overlook. From here, on a clear day you can see Catalina Island, and to its right, small pointy Santa Barbara Island. Farther to the right, low and flat San Nicholas Island is sometimes visible nearly 70 miles away. Looking north you can see the rock-ribbed slopes of Calabasas Peak. To the east are the towers of Century City and downtown L.A., and beyond, Mount Baldy, the San Bernardino Mountains, and Mount San Jacinto a good 100 miles off. To the southeast are the Santa Ana Mountains in Orange County.

From here, you can return the way you came or continue westward on to Segment 8 by turning left at the bottom of the summit spur trail.

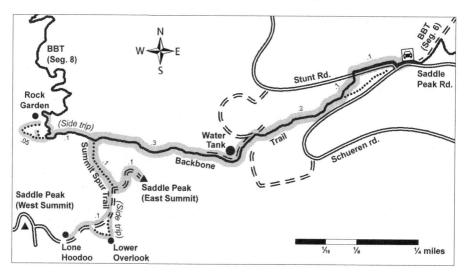

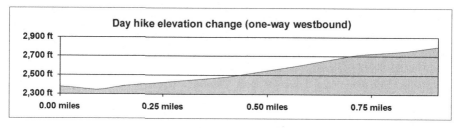

GPS COORDINATES

1. N34° 4.878', W118° 38.734'    2. N34° 4.850', W118° 38.829'    3. N34° 4.814', W118° 38.863'
4. N34° 4.750', W118° 39.050'    5. N34° 4.738', W118° 39.393'    6. N34° 4.653', W118° 39.352'
7. N34° 4.599', W118° 39.367'

# SADDLE PEAK TO STUNT ROAD CUTOFF

## ("SADDLE PEAK TRAIL" WEST APPROACH)

BBT segment length: 1.2 miles
Day hike length: 1.6 miles one-way
Suggested day hike: west to east, out-and-back

Elevation gain westbound: none
Elevation gain eastbound (to summit): 875 feet
Difficulty: moderate – Shade factor: 35%

Bikes not allowed – Horses not advised – Dogs not allowed

Access: N34° 5.156', W118° 39.610'

Saddle Peak, at 2,805 feet, is one of the highest points in the Santa Monica Mountains, and its close proximity to the ocean guarantees one of the most spectacular views from anywhere in the range – a breathtaking vista of coastline from Malibu to Orange County.

This approach from the inland side of the mountain is the tougher of the two routes to the summit, but it's also more rewarding. There's nothing like the experience of climbing up a semi-arid mountainside in an environment that appears to be somewhere far inland, only to come over a pass and suddenly see the Pacific Ocean nearly 3,000 feet below you, spreading out in a jaw-dropping panorama of indigo blue. Even after years of hiking in California, this is still thrilling to us, and it's the kind of experience that the Backbone Trail delivers best.

This hike feels more remote and wild than the easier eastern approach, and it delivers a few other surprises along the way. One of them is a spectacular "rock garden" full of pinnacles, crags and honeycombed hoodoos, a few of which have a pink tint to them reminiscent of Arizona or Utah.

Another is the abundance of manzanita. With its beautiful smooth red and brown bark standing out amidst the greenery, the manzanita is the celebrity of the chaparral plant community, and this segment has more of it than any other... so much in fact that at times it resembles a manzanita forest.

Other delights along this hike include unexpected sections of shady

*Gateway Rock looms like a fortress over much of the hike*

woods and panoramas of the upper Conejo Valley, the craggy Goat Buttes of Malibu Creek State Park, and much of the San Fernando Valley. The only fee for this fine scenery is a seemingly endless series of switchbacks.

Since this is a "destination" hike to a peak, there is no car access at the end of the segment. From there, you'll need to turn around and hike back another 1.6 miles or follow through on to adjacent Segment 7.

**Directions to trailhead:** From Pacific Coast Highway, take Malibu Canyon Road north for 6.3 miles, then turn right onto Mulholland Highway. Take Mulholland for four miles, then turn right onto Stunt Road. (If you're coming from the 101 freeway, exit Mulholland Drive in Calabasas and take it south for .6 miles. Turn right on Valmar Road – aka Old Topanga Canyon Road – and take it 1.1 miles. Turn right on Mulholland Highway, go another 3.9 miles, then turn left on Stunt Road.)

Once you're on Stunt Road, take it up the mountain for 2.8 miles. Pass mile marker 2.83, then park in any of the gravel turnouts along the north side of the road. Look for the trailhead[1] on the south side of Stunt Road just a few steps downhill from mile marker 2.92. A sign here reads BACKBONE TRAIL TO PIUMA ROAD 3.2 MILES.

**The hike:** Start off on the path – not actually the Backbone Trail but a connector path created to assist trail construction in the 1980s – which climbs a sandy hillside through sun-exposed low chaparral and connects to the Backbone Trail in about .2 miles.[2]

At the junction with the Backbone Trail, a post is marked 3 MILES TO SADDLE PEAK; .3 MILES TO STUNT ROAD. Turn left onto The Backbone Trail. (A right would take you via Segment 9 all the way down the slopes of Saddle Peak to Piuma Road.)

The trail starts off by heading gently uphill along a chaparral-covered shelf, paralleling Stunt Road which is down below to the left. This section offers some good panoramas of the upper Conejo Valley as well as Calabasas Peak on the right.

After about five minutes, the trail finishes contouring along the shelf. It crosses a little grassy flat, then enters a mix of chaparral and oak woods. The elfin forest of manzanita, red shank, ceanothus and chamise thickens enough to create a tree tunnel.

Around .6 miles, the trail turns south and begins a ⅔-mile-long series of tight switchbacks up the side of the mountain. On the map these switchbacks seem curiously confined to an extremely narrow range. Why is this? Because at the time of the trail's construction, the strip of land owned by the

*Sunset rays illuminate the Rock Garden*

National Recreation Area was very thin and boxed in on both sides by private property. The trail had to stay within the corridor regardless of the lay of the land.

The switchbacks seem to go on forever, and after a while you may feel like you've become caught in a Twilight Zone-esque time loop, during which you keep repeating the same stretch of trail every two minutes. What appears to be the same chaparral keeps cycling from short to tall, from sunny to shady and back again – as do the same ruts and frequent sandy patches along the trail.

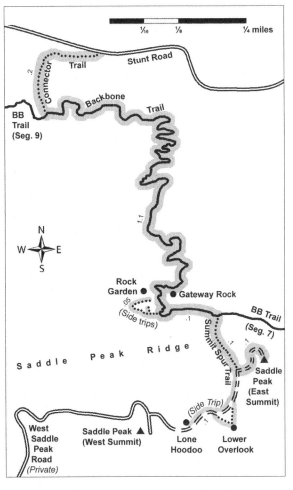

The sole break from the trail's monotony here is the beautiful manzanita which, along with ceanothus, is abundant throughout these switchbacks. The manzanita's bark is mottled with smooth red and brown sections, making it look to us like its trunk has been dipped in molten chocolate. As you climb, the sporadic sounds from Stunt Road down below drift away. Expect some rocky parts on this stretch, necessitating hiking boots.

At the top of the tightest stretch of switchbacks (about .8 miles), the chaparral lowers to unfold a great panorama of the Conejo Valley. Looking to the northwest, one can easily make out the Goat Buttes in Malibu Creek State Park, the flat-topped ridge of Ladyface Mountain, and layer after layer of hazy ridges which seem to go on forever. To the northeast, the sprawl of much of the San Fernando Valley is visible.

As you climb, eventually you'll notice a huge, block-shaped rock up ahead, which looks like a small fortress on the mountain. The closer you get to this stone monolith, known as Gateway Rock, the more impressive it looks.

Shortly before you pass below Gateway Rock, the sun-shielded chaparral grows taller, allowing for some in-and-out bits of shade, eventually becoming an elfin forest. Plenty of moss begins to appear along the trail, clinging to the rocks scattered about.

The trail passes just below the monolith at 1.2 miles, and soon after, you'll find that Gateway Rock actually does form a sort-of gateway to a fun surprise hidden high on the mountain. It's an area known as the Rock Garden, a cliff-ringed patch of greenery strewn with colorful sandstone rock pinnacles, many honeycombed with cavities. Tall grass and an abundance of manzanita mixed with ceanothus, toyon, and chamise help populate the "garden."

The trail climbs right through the middle of the boulder-strewn pass, becoming quite steep, rocky and sun-exposed, with a couple of narrow passages. A series of picturesque stone steps helps ease the climb.

Near the top of the Rock Garden, at about 1.3 miles, the Backbone Trail turns sharply left at a "T"-shaped junction[3] and heads along the base of a long rock slab.

*Highliners at the Corpse Wall in the Rock Garden*

**SIDE TRIP:** *A detour from the T-junction will take you down into the bottom of the Rock Garden, a cozy little maze of rocks and manzanita where the tan, pink and maroon pinnacles make it feel like you've been transported to some lovely hideaway in Arizona or Utah. It will also take you to the top of "the Corpse Wall," an impressive cliff (named for a dead body discovered at its base), which is popular with climbers.*

*Turn right onto the small path that runs along the base of the rock slab. A few seconds of walking will bring you to a minor junction.[4] The tiny path to the right heads up to the top of the Corpse Wall in about a minute. The cliff edge offers a great view looking down into the Rock Garden and over the Conejo and San Fernando Valleys. From here you may be surprised to see a handful of homes atop the mountain, accessed by a private road over the west summit. Please exercise caution on the tilted, exposed ledges at the Corpse Wall's sheer cliff; use your own judgement and venture here at your own risk.*

*To access the bottom of the Rock Garden, continue straight at the minor junction instead of turning right. Follow the path across a bit of rock slab, then continue straight at another minor junction,[5] about .02 miles from where you left the Backbone Trail.*

*From the junction, the path descends behind the back of the Corpse Wall rock slab. You'll come to a small Y-junction[6] at .03 miles. Follow the path to the right, which heads through a little gap in the slab, then works its way down into the Rock Garden.*

*The path winds its way eastward, doubling back in the direction of the Backbone Trail, below the Corpse Wall and past colorful pinnacles and boulders. It dead-ends at .05 miles, just below the Backbone Trail, which is at the top of a steep slope that's best not to attempt. Turn around and retrace your route to the Backbone Trail from here.*

After you exit out the top of the rocks, stupendous views open up on the left of the upper Conejo Valley. You're now looking *down* on Calabasas Peak, one of the easiest mountains in the Santa Monica Range to spot due to the conspicuous series of rock ribs angling up its side. Behind it and to the right are the towers of Warner Center, the tallest buildings in the western San Fernando Valley. Farther away to the north is the long ridge of 5,760-foot Liebre Mountain.

In another minute of walking, a flat rock sitting right in the middle of the trail looks like it would make for a good picnic spot with a spectacular view looking out over the Conejo Valley. But if you are thinking about stopping for lunch, you might want to press on for another ten minutes to a spot we call the "Lower Overlook", where a picnic spot with a million-dollar ocean view awaits.

At 1.4 miles, you'll come to a junction[7] where a sign marks the state park boundary. From here the Backbone Trail turns left onto Segment 7, descending gently toward the Lois Ewen Overlook (see previous chapter). To access Saddle Peak's summit, turn right onto the spur trail. (Even if the peak isn't your destination, the views from the top make it worth the short detour.) The spur trail climbs about 100 feet at a moderate pace, taking five to ten minutes to reach the east summit.

As you cross the "saddle" of Saddle Peak, the first view of endless blue ocean is a shock to the senses after climbing so much inland mountain terrain. As you look across the saddle, you'll notice a bizarre hoodoo (a rock tower) right in the middle of the dip and on the rim of the 2,700-foot drop towards the Pacific. From afar it nearly resembles the time-ravaged ruins of an ancient watch tower high above the sea.

*Looking down on Calabasas Peak from the BBT above Gateway Rock*

Here, at 1.5 miles, you will reach another junction.[8] The old road heading to the right will take you down to a lower overlook and to the lone hoodoo.

**SIDE TRIP:** *To reach the overlook and hoodoo, take the right fork and walk downhill along the old road. In about a minute, a path heads to the left[9] and up to the overlook.*

*From the overlook you'll enjoy views of the ocean, the city, Palos Verdes, Catalina Island, and a different perspective than the summit offers looking sharply down into the canyons below the peak. To the west, you'll get a glimpse of waves breaking at Malibu Lagoon. A series of smooth boulders along the rim of the overlook (both to the left and the right) offer several good sitting spots, perfect for picnicking or just hanging out.*

*After heading back down to the dirt road, if you continue farther along the road for another minute, you'll get to the lone hoodoo, which is interesting but not a must-see. The hoodoo actually looks more impressive from across the saddle than from up close.*

*The road continues up to the west summit, but the top is off-limits, precluding any westward views into Malibu Canyon, and it isn't worth the effort.*

To reach the east summit, turn left at the junction and continue uphill. In a few minutes, the old road ends atop the summit, which is windswept, treeless, and devoid of structures. You've come a total of 1.6 miles, but it may seem a bit farther.

From the top you'll enjoy an unforgettable view of miles of coastline, including Catalina Island, and to its right, small pointy Santa Barbara Island. Further to the right, flat San Nicholas Island is sometimes visible nearly 70 miles away. Looking north, the most obvious landmark is the rock spine tilting up Calabasas Peak. To the east lie the towers of Century City and downtown Los Angeles. Beyond lie Mount Baldy, the San Bernardino Mountains, Mount San Jacinto 100 miles away, and the Santa Ana Mountains in Orange County.

*Steps on the Backbone Trail in the Rock Garden*

From here, you can return the way you came or turn right at the bottom of the summit spur trail to head eastward onto Segment 7 of the Backbone Trail.

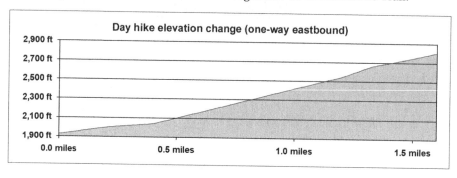

GPS COORDINATES
1. N34° 5.156', W118° 39.610'
2. N34° 5.070', W118° 39.732'
3. N34° 4.747', W118° 39.506'
4. N34° 4.744', W118° 39.512'
5. N34° 4.747', W118° 39.525'
6. N34° 4.746', W118° 39.544'
7. N34° 4.738', W118° 39.393'
8. N34° 4.653', W118° 39.352'
9. N34° 4.599', W118° 39.367'

## Segment 9:
# Stunt Road Cutoff to Piuma Road
## ("Saddle Creek Trail")

BBT segment length: 3.2 miles
Day hike length: 3.4 miles one-way
Suggested day hike: east to west, out-and-back or one-way

Elevation gain westbound: 214 feet
Elevation gain eastbound: 1,342 feet
Difficulty: moderate – Shade factor: 35%

Bikes not allowed – Horses allowed – Dogs not allowed

East access: N34° 5.156', W118° 39.610'     West access: N34° 4.571', W118° 41.167'

This hike, one of the finest stretches of the Backbone Trail, takes you on a trek through many different environments as it slowly rambles its way down the slopes of massive Saddle Peak. Along the way you'll encounter deep and moist forests, mountain streams, a secluded canyon under huge rock outcroppings, saddles ringed with lovely meadows, and an open ridge with sharp drop-offs overlooking the pastoral Conejo Valley. You'll also find plenty of something you don't typically associate with Southern California: moss. One thing you won't see, however, is the ocean.

Somewhere in the middle of this hike it may hit you: "Wow, I really feel like I'm in the middle of nowhere!" In Saddle Peak's remote upland canyons, you will be completely free from any signs of civilization – no buildings, no powerlines, no roads – only the great northern shoulder of the mountain with its seemingly endless ribs of rock watching over your own secret patch of wilderness. And all this just five miles from the hustle-and-bustle of The San Fernando Valley.

This hike could be done either as a long out-and-back trip from Stunt Road or as a moderate one-way hike of a few hours, mostly downhill, starting at Stunt Road and ending at Piuma Road.

*Monte Nido sits below the Backbone Trail in a bowl surrounded by peaks*

**Directions to trailhead (if hiking out-and-back):** From Pacific Coast Highway, turn right onto Malibu Canyon Road and follow it north for 6.3 miles, then turn right onto Mulholland Highway. Follow Mulholland for another four miles, then take a right onto Stunt Road. (If you're coming from the 101 freeway, exit in Calabasas at Mulholland Drive and take it south for .6 miles. Then turn right on Valmar Road – alternately known as Old Topanga Canyon Road – and continue for 1.1 miles. Then take a right onto Mulholland Highway. Follow Mulholland for another 3.9 miles, then turn left onto Stunt Road.)

Once you're on Stunt Road, take it up the mountain for 2.8 miles. Pass mile marker 2.83, then park in any of the gravel turnouts along the north side of the road. Look for the trailhead[1] on the south side of Stunt Road just a few steps downhill from mile marker 2.92. A sign at the trailhead reads BACKBONE TRAIL TO PIUMA ROAD 3.2 MILES.

*Alder and sycamore woods in Dark Canyon*

**Directions to trailheads (if using two cars for a one-way hike):** First, park your pick-up vehicle at the end of the hike. (Note: this location is also where you'd park if you use a rideshare service to shuttle you to the start of the hike.) Coming from Pacific Coast Highway, take Malibu Canyon Road north for 4.6 miles, then turn right onto Piuma Road. (If you're coming from the 101 freeway, exit at Las Virgenes Road and take it south for 5.1 miles, then turn left onto Piuma Road.)

Follow Piuma Road up the mountain for about a mile, then park near the trail terminus,[2] which is at the sharp right bend in the road across from mile marker 1.19. Parking along Piuma Road is somewhat limited, but the best bets are two gravel turnouts, both on the south side of the road. The first is downhill from the trail at mile marker 1.07, while the second is uphill from it at mile marker 1.26. A few other tiny turnouts may offer limited supplemental parking along the road as well.

Next, drive your drop-off vehicle (or rideshare) to the start of the hike. To do this, continue uphill on Piuma Road, eventually crossing over a pass with views of the ocean. 5.4 miles from where you dropped your pick-up vehicle, turn left onto

Schueren Road. Take Schueren Road for 1.8 miles to the top of another pass, then make a sharp left onto Stunt Road. Drive downhill on Stunt Road for another 1.2 miles, then park in any of the many gravel turnouts along the north side of the road near mile marker 2.92.

Look for the trailhead[1] on the south side of Stunt Road just a few steps downhill from mile marker 2.92. A sign here reads Backbone Trail to Piuma Road 3.2 Miles.

**The hike:** From the Stunt Road trailhead,[1] start off on the path (not actually the Backbone Trail but a connector path created to assist trail construction in the 1980s), which climbs a sandy hillside through sun-exposed, waist-high chaparral. You'll connect to the Backbone Trail at a "T"-junction in about .2 miles.[3]

At the Backbone Trail junction, a post is marked 3 Miles to Saddle Peak; .3 Miles to Stunt Road. Turn right, heading into the thick woods. (Going left would take you along Segment 8 up the nearly 1,000-foot climb to the top of Saddle Peak.)

For roughly the next quarter mile, the trail descends gently along the north slope of the ridge, winding in and out of small rivulets under the shade of a cool and dark oak and bay woodland. During the rainy season it can be wonderfully moist back in these woods, reminiscent of the Appalachians. Leaves cover the ground and moss is plentiful. In some spots, the tangled branches of toyon, bay, ceanothus and hollyleaf cherry seem jungle-like, as if somebody had to hack the route with a machete.

*Natural stone face along the trail*

At .4 miles, the trail exits the forest for an area of tall chaparral, providing occasional shade and sporting intermittent patches of moss and plenty of beautiful, red-barked manzanita. You'll get occasional views to your right looking out over Cold Creek Canyon and the much larger Conejo Valley beyond, with the Simi Hills forming a backdrop. In the far distance stand the peaks of Mount Pinos and Frazier Mountain, both over 8,000 feet. Down the mountainside, the hairpin curve of Stunt Road forms an obvious landmark. Eventually, rock-ribbed Calabasas Peak comes into view over your right shoulder.

At ½ mile, you'll reach a small crest that offers a view looking straight ahead over a little knoll (which you will be passing shortly) and to several farther ridges, including the Goat Buttes of Malibu Creek State Park. Just beyond the crest, a funky looking sandstone boulder sporting yellow and pink striations overhangs the trail, looking like an enormous piece of sponge cake as you approach it, but resembling the profile of a balding, big-nosed face when viewed looking back from the trail just past it.

At .8 miles, after a single switchback, you'll cross seasonal Saddle Creek in a wide, sweeping, low canyon below a series of tilted ribs of white sandstone that stretch far up the mountain. The stream has polished the rocks above the trail into a series of little smooth flumes. About a minute later, the trail travels along a streambed for a few steps. Head up the streambed and make a right after a big pink boulder.

For the next few minutes, the trail crosses a few minor streams and winds around occasional patches of boulders. At .9 miles, you'll pass by a series of enormous rocks, some nearly completely covered in moss, which look like they came crashing down the mountain eons ago.

The trail crosses over a small saddle between two canyons at 1.1 miles, with the white rocks of Saddle Peak forming a panoramic backdrop to the south. It then descends through an elfin forest of ceanothus and laurel sumac, seasonally lined with tiny ferns, grass and clover. Soon after, it skirts the edge of a hidden mountain meadow at a lower pass, watched over by dramatic sandstone outcroppings that go almost all the way up the mountain. A puny footpath[4] at the meadow (1.2 miles) heads off to the right but goes nowhere and dissipates into the brush.

In this uninhabited upland country, you'll feel far away from it all, but although the rambling nature of this trail might make you think it is comprised of an ancient Tongva or Chumash footpath, it's actually only about thirty years old. It was built in 1987 by the California Conservation Corps and L.A. Conservation Corps, which provide work experience for young people. Everyone who enjoys this trail owes them a debt of gratitude.

After crossing the meadow, the trail descends an isolated canyon that cuts into Saddle Peak's northern flank. You'll begin to get views of both Dark Canyon and Malibu Canyon straight ahead. During the rainy season you might be able to hear and see a miniature waterfall as it cascades into the canyon on your left at 1.3 miles.

The trail continues to steepen as it descends farther down the sun-exposed, sage-covered slopes of the canyon. It completes a single large switchback and, at 1.6 miles, crosses the stream at the bottom of the canyon.

The trail climbs briefly up the canyon's south slope. At 1.8 miles, it traverses a saddle between two canyons just below a peaklet of about 1,600 feet. Here it crosses through another mountain meadow of grass, sage and sagebrush, which affords fine views in both directions. The view to the right looks down on the Las Virgenes area, across the Conejo and Simi Valleys, and all the way to the distant high peaks of Ventura County and even the southern Tehachapi Mountains on a very clear day.

After a few minutes of panoramas, you'll leave the mountain meadow and plunge into a cool stretch along the sharply dropping north shoulder of the peaklet, shaded by toyon and ceanothus and lush with ferns and moss. This cute patch of miniature woods, the last bit of real shade you'll enjoy until Dark Canyon, continues for about five minutes.

*Crossing the saddle at 1.1 miles*

The trail then begins descending the ridge through a more open section. The drop-off here is sharp and the panoramas looking north across the valley to the high peaks in the distance and over the hills to Simi Valley on the right are unobstructed.

At 2.1 miles, when the trail turns sharp right, you'll come to the top of a major series of switchbacks – probably the toughest set of switchbacks on the entire Backbone Trail. There are about twenty of them, and they seem to go on forever as they wind 700 feet down the mountain. Although other segments may have more tightly winding switchbacks (most notably the Etz Meloy set in Segment 17 and the Saddle Peak set in Segment 8), this set has a heftier elevation change and feels more intense than those others.

That being said, they're still not that bad. Most of this mile-long circuitous stretch is only moderately steep and makes for a reasonably pleasant hike if the sun's not too strong. You'll get only very limited shade as you descend the slope through a mix of scrub oak, laurel sumac, ceanothus, mountain mahogany, sage, chamise and even some yucca. This section can be particularly charming in the spring when the path is lined with all kinds of wildflowers. Expect the trail to have a decent number of loose rocks on it, requiring proper hiking footwear.

About two minutes down the switchbacks, as the trail bends gently to the left, you'll get the best of what will prove to be numerous vistas looking sharply down on Monte Nido, a small semi-rural community that every year seems to have a few more houses in it. The hamlet lies in a picturesque setting, situated in a bowl ringed by chaparral-covered mountains. To the left, Piuma Road snakes its way up Piuma Ridge. Beyond Monte Nido is the sharp-topped peak of Brents Mountain. A little farther to the right, in the distance beyond the rolling Las Virgenes country-side, you might be able to make out the lofty ridge of Pine Mountain in Ventura County and the 8,283-foot peak of Mount Abel to its right.

At 2.4 miles, you'll descend a particularly steep and rocky section that lasts only about three minutes before the trail resumes a more moderate descent. Later, an overgrown side path[5] heading back sharply to the right at 2½ miles leads steeply downhill to the community below but ends at a closed gate. Continue straight.

You'll receive a brief but welcome respite from the sun in a tiny shady nook as you cross over a dry cascade in a ravine at 2.7 miles. This morsel of deep shade only lasts a few seconds, so appreciate it before you resume switchbacking down the exposed ridge.

Three miles into the hike, a rinky-dink path[6] heads off sharply to the right and down to Hilltop Climb Drive in Monte Nido. Continue straight.

You will finally complete the switchbacks a few minutes later when the trail reaches a tributary of Dark Canyon. As the trail follows the tributary downhill, elfin forest begins to close in and thicken. Soon you'll hear the sound of water up ahead in Dark Canyon as the woods become full and shady.

The trail crosses the stream at the bottom of Dark Canyon at 3.2 miles. This is an aptly named place, with a classic, picture-perfect mountain brook tumbling over boulders, shaded by white alders, sycamores and bay trees, and laced with ferns, moss, and wild grape vines. Although the drought years have taken their toll, it's still the prettiest stretch of woods on the hike. It looks like the type of thing you'd expect to find at higher elevations than here at about 700 feet. Unfortunately, your visit is brief as the trail promptly heads up the other side of the canyon, climbing about 100 feet and leaving the riparian woods behind for more elfin forest.

A few minutes' walk beyond the crossing, you'll pass a little viewpoint looking down into Dark Canyon off to the right. Shortly after, at the top of

*Seasonal cascade just off the trail*

the climb, you'll reach at an unmarked junction[7] at 3.3 miles. Take the left fork downhill. Don't go straight and uphill, which soon leads down the ridge into a neighborhood.

From the junction, the trail descends a chaparral-covered draw to its terminus[2] at Piuma Road (3.4 miles). If you didn't park a pick-up car at the trailhead, you can return the way you came by heading back up the mountain. You can also continue on to Segment 10 of the Backbone Trail by crossing the road, walking along it for a few steps to the left, and picking up the trailhead on the right, signed BACKBONE TRAIL TO TAPIA PARK 2.1 MILES.

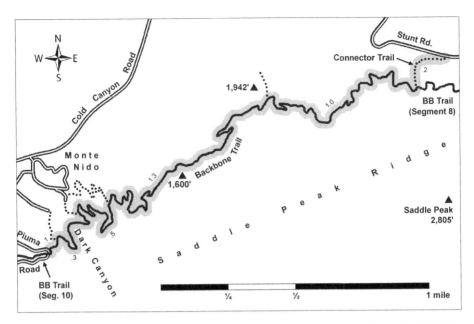

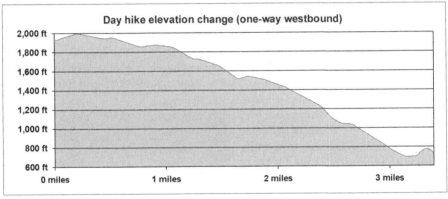

GPS COORDINATES

1. N34° 5.156', W118° 39.610'    2. N34° 4.571', W118° 41.167'    3. N34° 5.070', W118° 39.732'
4. N34° 5.034', W118° 40.304'    5. N34° 4.714', W118° 40.930'    6. N34° 4.683', W118° 41.043'
7. N34° 4.608', W118° 41.135'

# PIUMA ROAD TO LAS VIRGENES ROAD

## ("PIUMA RIDGE TRAIL")

**BBT segment length: 2.0 miles**
**Day hike length: 2.0 miles one-way**
**Suggested day hike: west to east, out-and-back**

**Elevation gain westbound: 225 feet**
**Elevation gain eastbound: 468 feet**
**Difficulty: easy to moderate – Shade factor: 75%**

**Bikes not allowed – Horses allowed – Dogs not allowed**

**East access: N34° 4.562', W118° 41.170'      West access: N34° 4.798', W118° 42.230'**

This short and relatively easy hike offers no ocean views, no dramatic cliffs, no spectacular summits. So why go here? Because it will take you on a delightful, gentle walk through some of the best forests that the Backbone Trail has to offer, particularly a few brief but wonderful patches of deeply shaded bay woods through which the trail twists and turns amongst tilted boulders, tangled roots, moss, ferns and little hollows. If you're walking here near dusk, it can get just a bit spooky, as if gnomes are about to pop out from behind the rocks or from the hollows of trees.

In fact, this hike offers a sampling of *four* different types of local forests. You'll start and finish in riparian woodlands, traverse a lovely coast live oak forest, cross through bay woods in the middle, and pass a fine mix of elfin forest in between.

In spite of the oak groves and rock-garden woods, this is hardly wilderness; during much of the hike you can hear distant noise from both Piuma Road and Malibu Canyon Road, as well as from the community below. You're never far from civilization, and you're not all that high on the mountain either, but you will enjoy pastoral views looking over the small village of Monte Nido framed by the Goat Buttes of Malibu Creek State Park, as well as rugged Saddle Peak in the other direction. You'll also have the option to cross beautiful Malibu Creek, which almost always has water in it.

This hike, one of the shadiest BBT segments, can be done in either direction; we describe it eastbound solely because there is ampler parking at its western end.

*Malibu Creek crossing*

*Saddle Peak seen from the high point on the trail*

**Directions to trailhead:** From Pacific Coast Highway, take Malibu Canyon Road (aka Las Virgenes Road) north for 4.5 miles. About a minute after exiting the tunnel, and shortly before the traffic light intersection with Piuma Road, park in the Las Virgenes Road fee-based lot on the left.[1]

If you're coming from the 101 freeway, exit Las Virgenes Road and take it south for 5.1 miles. Shortly after you pass the intersection with Piuma Road, turn right and park in the Las Virgenes Road fee-based parking lot.[1]

The parking lot has restrooms but no drinking water. On rare occasions when the lot is full, there is additional fee-based parking at Tapia Park (a day-use division of Malibu Creek State Park), which is less than half a mile north on Las Virgenes Road. (Tapia Park also has flush-toilet restrooms and drinking water.)

**The hike:** As soon as you exit your car, you have a choice to make. You can start off immediately on the Backbone Trail, which includes a foot crossing of Malibu Creek, or you can bypass the creek via an alternate route over the bridge on busy Las Virgenes Road, reconnecting with the Backbone Trail a few minutes later.

Crossing the creek is far more picturesque and it's technically the correct route – satisfying for Backbone Trail purists like us. Plus, you'll have the luxury of taking the trail *under* busy Las Virgenes Road instead of dashing across it like a squirrel, but there's a catch. Be forewarned that Malibu Creek is by far the most substantial water crossing on the entire Backbone Trail, and at times it has a healthy amount of water in it. You could soak your feet and more during wetter periods. This may seem like an unnecessary risk to take when you have a large bridge just yards away.

On the other hand, if you take the alternate route across the bridge, you will have to scurry across Las Virgenes Road without a light or crosswalk as cars come at you in excess of 50 mph. Then you must walk along the shoulder of the busy road. Not the most pleasant start to a hike, but at least your feet will stay dry.

If you have a few minutes to spare, we suggest trying the trail to the creek crossing. When you get there, if the water looks too high or if you're unsure, play it safe and retrace your steps back to Las Virgenes Road to cross the bridge instead.

To take the Backbone Trail via the creek crossing, start off on the short connector trail that leaves from the center of the parking lot where a sign reads BACKBONE TRAIL.[1] Within half a minute you'll pass a small bathroom on your left. Just beyond it, the connector trail reaches a junction[2] with the Backbone Trail. A left turn here (up the railroad-tie steps) would take you westbound on Segment 11; go eastbound instead by continuing straight.

In a minute there's another little junction;[3] this one is not signed. Take the smaller path to the right, which heads steeply downhill. (The path heading straight wanders towards the nearby Las Virgenes Water Treatment Plant but deteriorates into an overgrown bushwhack.)

In another minute, the path crosses a service road[4] for the water treatment plant. Continue straight across the road

*Cabin ruins just off the trail*

and soon (at roughly .15 miles) the dwindling path will lead you to the banks of Malibu Creek. Cross here using the stepping stones; don't take the trail to the left[5] that parallels the creek westward and quickly rejoins the service road.

Continue past the creek on the faint path, bearing left at a false T-junction[6] in a dry part of the streambed less than a minute past the main creek crossing. Roughly another minute later, the path joins a more substantial trail at another T-junction[7] by an old stone chimney. Turn right, paralleling the creek. (The short trail to the left leads to Tapia Park, where there are picnic tables, a drinking fountain and water spigot.)

Soon the path will take you through an underpass below busy Las Virgenes Road. It's a bit seedy under here (as underpasses tend to be), but at least you don't have to run across the road. Just beyond, the trail turns left, heads up a short hill, and emerges from the woods onto Piuma Road[8] at about .4 miles. Turn right and walk along the wide south shoulder of the road.

**ALTERNATE ROUTE (TO BYPASS CREEK CROSSING:** *From the parking lot, cross Las Virgenes Road and turn left, heading north along the east shoulder of the road. Cross the bridge over Malibu Creek via the pedestrian walkway.*

*In about four minutes' walk, Piuma Road intersects Las Virgenes Road at a traffic light.[9] Follow Piuma Road to the right along the wide south shoulder. Within another half minute, the Backbone Trail (the creek crossing route which you bypassed) emerges onto the road out of the woods on your right.[8]*

Continue eastbound along the wide south shoulder of lightly travelled Piuma Road. After a few minutes, you'll pass road mile marker 0.27. About a half minute beyond the marker, watch for a 30 MPH sign and another sign showing a car skidding on a wet road. Between those two signs, locate the single-track Backbone Trail as it heads off to the right (at .6 miles).[10] An adjacent small metal sign reading BACKBONE TRAIL is easily missed.

Follow the trail down into the woods and hop across Cold Creek, a little tributary of Malibu Creek. (The very beginning of this stretch is sometimes overgrown with tall grass, but after a minute or two, the trail conditions improve.) Continue gently uphill through a large patch of thistle and lush grass dotted with cottonwood trees.

Soon the trail turns left and heads briefly uphill into an oak woodland. It closely parallels a lengthy wooden fence, behind which are animal cages owned by the California Wildlife Center, including a "rattlesnake habitat."

You'll come upon a tiny paved road[11] at .7 miles, in an area of oaks with a handful of buildings. The road leads up to the California Wildlife Center, an organization dedicated to the rescue, rehabilitation and release of California's native wild creatures, and at times you may hear animal sounds coming from the center. Cross the road and the little gravel parking area, then continue straight on the trail, which exits the parking area at its far left corner.[12]

The trail heads up along a shelf that drops off sharply to the left. Piuma Road is almost directly below, and you can hear the sounds of civilization wafting up from the community of Monte Nido: kids laughing, dogs barking, perhaps a horse whinnying. You'll leave the woods behind for an alternating mix of tall chaparral, grassland, and the occasional full-grown oak.

The trail passes the ruins of a couple of old cabins at .9 miles (which are worth a minute of exploring) and within a few minutes it curves right and heads westward for a stretch, beginning its climb up the ridge. As you ascend, the sounds from the community below fade away as the oak woods grow deeper and shadier.

The trail turns back eastward across a little glen at 1.1 miles. Here you'll pass through the first of three lush bay and oak forests flanking the mountainside.

Within a couple of minutes, you'll be out of the forest and climbing gently through more chaparral with some beautiful manzanitas. Then, at about 1.2 miles, you'll enter the second and best of the forests. This one has sprung from the remnants of an old landslide. The trail

*Through the rocky forest*

twists and turns its way through a jumble of tumbled boulders with patches of moss and ferns peeking out from between the rocks. The bay trees here provide surprisingly thick shade, creating a dark, almost gloomy wood.

*Looking west over Monte Nido towards the Goat Buttes*

After leaving this delightful rock-garden forest, you'll cross a dry stream at 1.3 miles and start ascending a couple of intermediately steep switchbacks, with the chaparral growing taller as you proceed. At about 1½ miles, you'll finish climbing the switchbacks and enter a shade-providing elfin forest of ceanothus and chamise.

Soon you'll enter the last of the rockslide areas, another boulder-strewn wood with plenty of shady bay trees. You can easily confirm a bay tree by rubbing the leaves and giving your fingers a quick sniff. If it smells like you are cooking a stew, you've got a bay tree on your hands... literally! The canopy here is thick, complete with plenty of moss and ferns, but it only lasts a few minutes.

You'll emerge back into tall chaparral. A couple of grass-covered old land-slides along this stretch offer partial views looking back over Monte Nido.

At 1.6 miles, shortly after crossing under some power lines, you'll crest the high point on the hike. This isn't a particularly isolated stretch of trail; road noise is audible from Piuma Road, which is now *both above and below you*, as well as from Malibu Canyon Road farther below.

From the high point, the trail slowly winds its way downhill, rounding some small ravines with seasonal streams. It takes on a sunnier, drier character as it passes through thistle, hollyleaf cherry, and some stretches of wall-to-wall chamise. But there's still plenty of patchy shade from areas of tall chaparral, including the occasional tree tunnel. You'll be rewarded with frequent views of Saddle Peak ahead. Some extra rocky sections may make you grateful for your hiking boots.

**ALTERNATE EXIT:** *At 1.8 miles, a short side path to the right[13] provides an exit to Piuma Road, adjoining it at a horseshoe bend just downhill from mile marker 1.52.[14]*

Within a minute after the alternate exit, as the trail makes a wide right turn across an open area, you'll enjoy your best view on the hike, looking off sharply to the left over Monte Nido and the Conejo Valley countryside. From here you can see Malibu Creek State Park in the background with the Goat Buttes rising prominently. The view, which extends to the distant high country of Mount Pinos and Mount

Abel (8,847 and 8,287 feet respectively), is particularly impressive considering that you're not very high up. The tiny housing subdivision directly below is an odd-looking place, as if someone took a bunch of expensive suburban houses and stuck them in the country.

From here on, you'll be closely paralleling Piuma Road. As you descend, the old chaparral grows steadily taller, creating more tree tunnels. It transitions into a mature riparian woodland shortly before the trail exits onto Piuma Road at mile two.[15]

This is the end of the segment. From here you can turn around the way you came or continue on to Segment 9 (a long uphill climb which we describe west-bound, starting from the far end). To access Segment 9, do NOT follow the dirt road that leaves directly across from the end of this segment. Instead, walk left on Piuma Road for a few steps, then turn right onto the signed Backbone Trail.

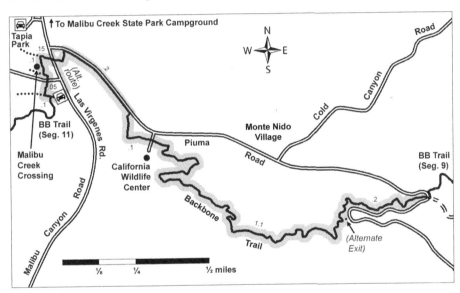

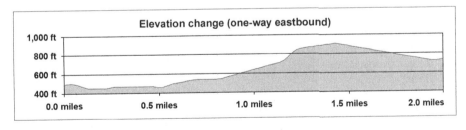

GPS COORDINATES

| | | |
|---|---|---|
| 1. N34° 4.798', W118° 42.230' | 2. N34° 4.812', W118° 42.255' | 3. N34° 4.816', W118° 42.291' |
| 4. N34° 4.852', W118° 42.288' | 5. N34° 4.865', W118° 42.300' | 6. N34° 4.894', W118° 42.298 |
| 7. N34° 4.915', W118° 42.303' | 8. N34° 4.921', W118° 42.230' | 9. N34° 4.924', W118° 42.264' |
| 10. N34° 4.771', W118° 42.034' | 11. N34° 4.707', W118° 41.990' | 12. N34° 4.706', W118° 41.965' |
| 13. N34° 4.512', W118° 41.419' | 14. N34° 4.515', W118° 41.398' | 15. N34° 4.562', W118° 41.170' |

# LAS VIRGENES ROAD TO CORRAL CANYON ROAD
## ("MESA PEAK MOTORWAY")

**BBT segment length: 5.2 miles**
**Day hike length: 5.2 miles one-way**
**Suggested day hike: west to east, one-way**

**Elevation gain westbound: 2,181 feet**
**Elevation gain eastbound: 677 feet**
**Difficulty: moderate – Shade factor: 10%**

**Bikes allowed – Horses allowed – Dogs not allowed**

**East access: N34° 4.798', W118° 42.230'     West access: N34° 4.914', W118° 45.496'**

If you've ever driven along the bottom of Malibu Canyon, looked up at its precipitous heights and thought, "I wonder what's up there," this hike is for you. The eastern part of this segment will take you along the canyon rim, with plenty of dizzying views looking nearly straight down on Malibu Canyon Road 1,500 feet below as it snakes its way toward the ocean.

The majority of the hike follows a dirt road, Mesa Peak Motorway, as it contours rather strictly along the top of a ridge through endless chaparral. So granted, parts of this hike can get a bit tedious… if by "tedious" you mean a nearly constant, stupendous ocean panorama that most hikers would beg for a mere glimpse of.

When you're not walking the rim of the canyon, you'll be treated to non-stop views of the Pacific shimmering 2,000 feet below, not to mention plenty of vistas of the Santa Monica Mountains and the beautiful rolling backcountry of Malibu Creek State Park, at one time the 20th Century Fox filming ranch. You'll be looking down on the shooting locales of countless movies, including the *Tarzan* films, *Planet of the Apes*, the famous jump into the river from *Butch Cassidy and the Sundance Kid*, and the *M*A*S*H* tele-

vision series. You'll also spend time crossing a spine of rock hoodoos, and, if you wish, walking a charming stone labyrinth.

What's the catch? In a word, perspiration. The reason for all these views is that there are virtually no trees to shade the dirt road as it heads through a semi-arid environment of low chaparral, yucca, scrub

*Backbone Trail crossing below McAuley Peak, 2,000 feet above the Pacific*

*Looking down 1,500 feet into Malibu Canyon from the trail along the rim*

and grass. Apart from a few brief patches of shade, you'll be under virtually constant sun exposure, so a hike in the morning or on a cool or cloudy day is advised.

The other challenge is the climb. If you were to hike westbound, you'd begin by trudging 1,500 feet up a relentlessly steep thigh buster, one of the toughest climbs on the Backbone Trail. Instead, we recommend hiking this segment one-way eastbound, requiring two cars (one at each end), a friend to shuttle you, or a somewhat pricey rideshare trip (see Chapter 3). This results in only 677 feet of total uphill gain; even then, coming down the escarpment can be tough on the calves.

**Directions to trailheads:** First, park your pick-up car at the end of the hike. (This is also where you'd park if you plan to use a rideshare service to shuttle you to the start of the hike.) From Pacific Coast Highway, take Malibu Canyon Road north for 4½ miles. About a minute past the tunnel, shortly before the intersection with Piuma Road, turn left and park in the Las Virgenes Road fee-based parking lot. (If you're coming from the 101 freeway, exit Las Virgenes Road and take it south for 5.1 miles. Shortly after you pass the intersection with Piuma Road, turn right and park in the Las Virgenes Road fee-based parking lot.)

Restrooms are located by the trailhead[1] (the only ones on the hike) but with no drinking water. On rare days when the lot is full, there is additional fee-based parking at Tapia Park (a day-use division of Malibu Creek State Park), less than half a mile north on Las Virgenes Road. Tapia also has flush toilets and drinking water.

Next, drive your drop-off car (or rideshare) to the start of the hike. Drive south on Malibu Canyon Road for 4½ miles to Pacific Coast Highway and turn right. Stay on PCH for 2.3 miles, then turn right on Corral Canyon Road. Take the road up the mountain for 5½ twisty miles until the pavement turns to gravel, then continue for another ¹/₁₀ mile and park in the large, dirt parking lot.

The Backbone Trail crosses the parking lot. This segment departs from the unsigned trailhead[2] located on the southeast corner of the lot (the same side from which you drove in).

**The hike:** The single-track trail begins by climbing a rock slab up the ridgeline into an area of tilted rock pinnacles, formed when slabs of Sespe Formation sandstone were left standing as the softer sandstone

*Brents Mountain viewed from the trail's descent into Malibu Canyon*

surrounding them eroded away. The pinnacles are complemented by a natural garden of manzanita, sage, chamise, yucca, and the occasional eucalyptus tree.

> **ALTERNATE ROUTE:** *If you'd like to bypass the somewhat rugged trail along the rock spine, you can take the dirt road around it and rejoin with the Backbone Trail in .6 miles. This route, while far less scenic, involves less climbing and no steep sections. It's also the mandatory route for bicyclists and horses.*
>
> *To get to the route, walk south along Corral Canyon Road (the road on which you drove in). After ¹/₁₀ mile, the road becomes paved. Continue for another .2 miles, then turn left onto dirt Mesa Peak Motorway.[3] In another .3 miles, the road rejoins the single-track Backbone Trail on the left.[4]*

A few minutes into the hike, you'll come to a "Y"-shaped junction[5] at .2 miles. The Backbone Trail angles to the left, staying below the ridgeline, while the smaller trail to the right heads up the ridge, closely hugging the base of the rock spine.

> **SIDE TRIP:** *The path to the right follows the top of the ridge for ¹/₁₀ mile, ending at the ruins of an old home situated amidst pinnacles. One wonders what the house must've been like, set alone in the mountains atop its little point, surrounded by rock towers.*

About a minute later, the trail passes the remains of an old vehicle – so badly crushed that it's unclear if it was a car or a truck – and then heads through a copse of oaks. Soon after, at the top of a climb up a short but very steep and rutted stretch, you'll pass some heavily graffitied water tanks on the right. These are the first of what will prove to be an almost comical number of dilapidated water tanks along this segment. From the tanks you get a fine view looking south over the ocean and a few lower ridges. The dirt road just below you is Mesa Peak Motorway, which the Backbone Trail will soon be merging up with.

*The labyrinth*

At .4 miles, the trail crosses a colorful stretch over the rim of a sandstone bowl flanked by more pinnacles. A short scramble down the steep trail will take you to where it joins dirt Mesa Peak Motorway[4] at the half-mile point.

**SIDE TRIP (CLOSED):** *A faint path leaving the trail at .6 miles[6] provided access – until recently – to Corral Canyon Cave, a quaint little rock grotto that was used by the indigenous Tongva people for religious practices. After a claustrophobic squeeze through an entrance slot colloquially known as "The Birth Canal," visitors could easily climb up into the cave, which becomes reasonably roomy and opens up to a ledge offering a fine view looking north over the valley. Unfortunately, the cave is now sealed off tight. Who's to blame for this? Well, in a way, Jim Morrison.*

*Several years ago, word spread that The Doors' singer used to hang out in the cave and write poetry. Although it's likely untrue, the rumor spread on social media, and soon fans began visiting the cave and partying there, leaving behind trash and graffiti. When the graffiti spread to the wider area, officials sealed off the cave.*

*The other cause for the closure was the increasing numbers of visitors who were toking up in the cave. The problem wasn't the pot, it was the fire. In 2007, cave partiers accidentally started the Corral Canyon Fire, which destroyed 53 homes and injured five firefighters, earning each of the partiers a year in prison. Now the Corral Canyon Arson Watch patrols this area and cracks down on anyone lighting a fire, including smoking.*

After the junction, a couple of minutes of easy walking will bring you to a flat area of gravel below a large sandstone butte pockmarked with cavities. Here, a spiral rock labyrinth beckons you to walk it. A short trail heads up through a gap in the butte into a fun area of small cliffs and tiny caves, good for a couple of minutes of exploring.

**SIDE TRIP:** *If you'd like to reach the top of the butte, the easiest way is to continue eastbound on the road past the butte, then make a sharp left onto a small use path[7] at .7 miles. The path heads up steeply, climbing the back side of the butte. It takes two or three minutes to reach the top, where you'll get a great view looking down on the labyrinth, as well as north over the Conejo Valley and south over Malibu.*

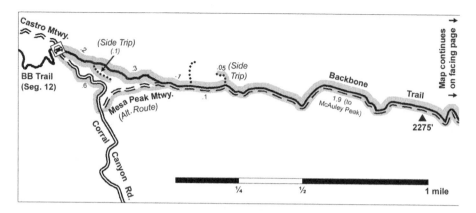

As you continue on the dirt road, the ocean vistas and sharp drop-off remain spectacular, with views of the Channel Islands and Malibu mansions far below. The occasional rock outcropping adds a little variety. At .8 miles, you'll pass one of these rocks (on your right just below the trail) that resembles an enormous skull.

The trail soon crests, then heads downhill at an almost imperceptible pace until it reaches a little saddle with a wide, flat section of road at 1.1 miles. The hills here create a small amphitheater, providing a sheltered home for numerous birds… as well as another rusted water tank.

(For the next few miles, a series of tiny use paths parallel the Backbone Trail, occasionally leaving and then rejoining as they contour up ridgelines. None are recommended as they are very steep, crumbly, hazardous and not worth the risk considering the great views available from the main route.)

From the saddle, the trail climbs steeply through a copse of old oaks, which burned in the 2018 Woolsey Fire. Several more minutes of climbing will take you over a crest just north of a small knoll (1²/₃ miles), where you're at the high point on the hike at 2,240 feet.

The crest offers a superb view looking north. Directly below are the craggy Goat Buttes and Udell Gorge carved by Malibu Creek (the setting for the aforementioned *Tarzan, Butch Cassidy* and *Planet of the Apes* films). Malibou Lake (sic) is to the left, and farther beyond, past the rolling Conejo Valley, are the high peaks of Mount Pinos and Frazier Mountain.

Looking straight ahead, Mulholland Highway winds past several rock cuts; the Santa Suzanna Mountains lie beyond. To the right, the green

*Descending the steep rim of the sandstone bowl*

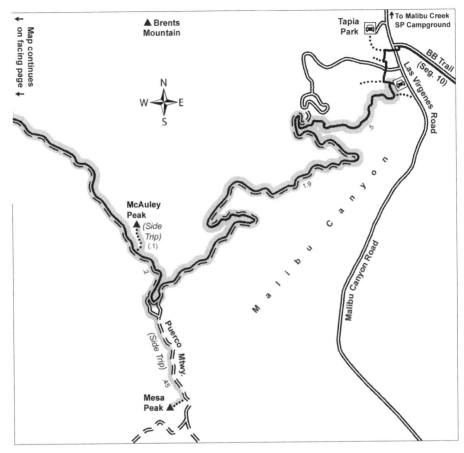

pastureland of the King Gillette Ranch stands out amongst the rolling knolls of the Las Virgenes area; the peaks below are the ones over which helicopters fly during the opening credits of M*A*S*H. Further right is the urban sprawl of the San Fernando Valley, with the high peaks of the San Gabriel Mountains beyond.

The trail then descends steeply, continuing its repetitive sun-exposed trek through low chaparral, yucca and wind-blown grass. At 1.7 miles, it zigzags down a short but conspicuous single switchback, which really is only notable because it can help in locating your position on a map.

About five minutes' walk past the bottom of the zigzag, where the road bends gently to the left at mile two, look for an impressive outcrop of pure white chalk up on the bluff to your left. When sunny, the vivid rock face completely stands out from its drab hillside, looking like someone had meticulously whitewashed it.

Five more minutes down the road, after another bend to the left, you'll enter a brief stretch where the roadbed has exposed hundreds of turritella fossils – the shells of ancient sea snails. These little corkscrew-shaped guys, about an inch long, populate the gravel underfoot but are easily missed if you're not looking down.

The fossils continue until just before your descent ends at a saddle (2.3 miles). From the saddle, the road steeply climbs the south side of a peak with a prominent boulder tilted on top. Until 2015, this landmark had the uninspired name of Peak 2049, but it was then officially renamed McAuley Peak for the hiking author and Backbone Trail advocate Milt McAuley, who died in 2008 (see Chapter 1).

At 2.6 miles, you'll pass just south of the peak, then over a slight saddle where you'll come upon – you guessed it – yet another rusted water tank.

> **SIDE TRIP:** *From the saddle, a faint side path[8] leads back up the ridge for about ¹⁄₁₀ mile to the summit of McAuley Peak and its tilted boulder. If you have extra time, the huge maroon-and-tan-striated boulder is a picturesque curiosity, but this side trip isn't a must because the boulder also obscures any chance of a 360° view from the top unless you scramble up it – a class III rock climb a bit beyond the scope of the average hiker.*

Soon you'll get your first (but far from best) view looking down into Malibu Canyon on your left. Descend to a "Y"-shaped junction[9] at 2.8 miles and follow the Backbone Trail to the left.

> **SIDE TRIP:** *Heading right at this junction will take you to Mesa Peak via a relatively easy walk of less than half a mile each way, along a dirt road with some up-and-down but little elevation change.*
>
> *From the peak you'll enjoy a 360° vista of the ocean, Malibu Canyon, the rocky slopes of Saddle Peak to the east, the San Fernando Valley and San Gabriel Mountains to the north, and the Backbone Trail winding its way atop the ridge you've just walked. Because Mesa Peak is closer to the ocean than any other spot on this hike, you'll see more detail of the coastline from here: Point Dume protruding into the ocean, the curve of beaches from Santa Monica to Redondo, the flat-topped hill of Palos Verdes Peninsula, and Catalina Island.*
>
> *To reach the peak, leave the Backbone Trail and head down the dirt road (called Puerco Motorway – most hikers are surprised to find that Mesa Peak Motorway does not actually go to Mesa Peak). ¹⁄₁₀ mile into the side trip, angle gently to the right at a junction.[10] (Turning sharp left here would take you back up to the Backbone Trail.)*
>
> *Pass an open chain-link fence at about .4 miles. A few steps beyond, a small use path heads to the right.[11] A brief scramble of about a minute up the gravelly and very steep path will land you at the summit, which is topped with low chaparral and a few metal poles.*

*Large, skull-like rock along the trail*

Just past the junction, a picnic table provides a lunch opportunity with a million-dollar view, including a glimpse of Point Dume protruding out into the sea. A few steps beyond, another trail heads off sharply to the right[12] and joins the previous side trail to Mesa Peak. Continue straight on the Backbone Trail.

Once past the junction, the trail abruptly turns northeast and begins its 1,400-foot descent along the rim of Malibu Canyon. As you descend for the next half mile, expect increasingly spectacular vistas looking down 1,500 feet into the canyon. Far below, Malibu Canyon Road slithers like a concrete serpent as scores of ant-sized drivers zoom north and south, oblivious that anyone is watching them from above. The Pacific lies beyond. The highest peak forming the opposite side of the canyon (with the communications towers atop) is Saddle Peak, accessible from Segments 7 and 8 of the Backbone Trail.

Malibu Canyon is a unique place. First of all, it's the only canyon that cuts *through* the Santa Monica Mountain Range instead of just cutting into one side of it like Topanga, Zuma or Big Sycamore Canyons. Also, it bisects the range almost exactly in the middle, separating the eastern half from the western half. And did you know that Malibu Creek was flowing along this same course *before* the mountains were even here? As the range slowly lifted, the creek just kept shoving its way through, carving the canyon.

This unforgettable stretch is augmented by some fine views looking out over the rolling Las Virgenes countryside to your left. Oaks along the road become plentiful but still don't provide any shade.

At 3.4 miles, the trail makes a sharp left turn down a large switchback as it heads away from Malibu Canyon and descends a smaller canyon on the north side of the ridge. Here the hike dramatically changes character: majestic oaks overhang the trail, rock formations loom above, tall chaparral and grass line the route, and there are even patches of reeds and moss. This cool respite is a treat to the senses after such a long trek along the dry ridgeline – not a bad place for a shady break. The vistas of Malibu Canyon are temporarily gone, but instead you'll enjoy gorgeous views looking out over the rugged and rolling backcountry of Malibu Creek State Park, with the sharp double-topped peak of Brents Mountain to the near northwest.

*Turritella fossils found along the trail.*

The trail exits the woods to revisit the rim of Malibu Canyon at mile four, where you'll find more great views and – hoo ha! – another rusted water tank.

Continue steeply downhill for another ¼ mile, with plenty of Malibu Canyon views, until the trail curves away from the rim again, this time for good. As before, it wiggles down the back of the ridge past more oaks, with similar views of the Malibu Creek area. The Las Virgenes Water Treatment Plant, an obvious landmark, soon comes into view down below.

Look for an easily missed junction[13] just after the road makes a sharp right hairpin curve at 4.7 miles. Here the single-track Backbone Trail branches away from the road, heading off to the right as the road curves back to the left. A sign reads BACKBONE TRAIL TO PIUMA PARKING LOT .6 MILES. (The road continues down to the water treatment plant but is promptly blocked by a gate.)

This single-track section of the Backbone Trail is rather picturesque, and a whole different world from that of the dirt road. It winds in and out of elfin forest, past oaks and a variety of other trees, through perennially green grass and under welcome shade, staying just out of sight of the water treatment plant down below.

After a few minutes of gentle climbing, the route begins to descend toward Las Virgenes Road down a steep and rocky pitch, the last thing your already battered knees will want.

At 5.2 miles, the trail forks at a junction[14] just above Las Virgenes Road. Follow the Backbone Trail to the left. (The overgrown right fork heads down to Las Virgenes Road, missing the parking area.) Head steeply downhill through a deep rut for a minute or two, finishing your descent via some wooden steps.

At the bottom of the hill, you'll come to a "T"-shaped junction.[15] The Backbone Trail turns left, heading off onto Segment 10 towards Piuma Ridge and Saddle Peak. To reach the parking lot and the end of this segment, turn right onto the spur trail. Pass the bathroom on your right (or use it – that's your business). A minute later, at 5.2 miles, you will be back at the parking lot.[1]

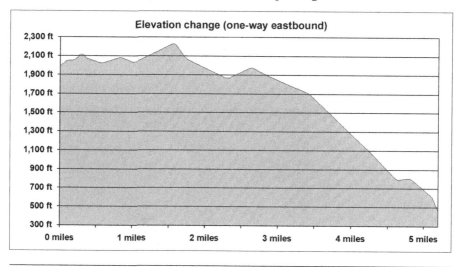

GPS COORDINATES

| | | |
|---|---|---|
| 1. N34° 4.798', W118° 42.230' | 2. N34° 4.914', W118° 45.496' | 3. N34° 4.742', W118° 45.332' |
| 4. N34° 4.813', W118° 45.036' | 5. N34° 4.861', W118° 45.364' | 6. N34° 4.807', W118° 44.957' |
| 7. N34° 4.808', W118° 44.818' | 8. N34° 4.252', W118° 43.337' | 9. N34° 4.107', W118° 43.280' |
| 10. N34° 4.043', W118° 43.248' | 11. N34° 3.778', W118° 43.144' | 12. N34° 4.099', W118° 43.243' |
| 13. N34° 4.682', W118° 42.664' | 14. N34° 4.772', W118° 42.255' | 15. N34° 4.812', W118° 42.255' |

## SEGMENT 12:
# CORRAL CANYON ROAD TO LATIGO CANYON ROAD
## ("CASTRO CREST")

**BBT segment length: 4.1 miles**
**Day hike length: 4.1 miles one-way**
**Suggested day hike: east to west, out-and-back or one-way**

**Elevation gain westbound: 946 feet**
**Elevation gain eastbound: 895 feet**
**Difficulty: moderate – Shade factor: 15% (45% pre-fire)**

**Bikes allowed – Horses allowed – Dogs allowed on leash**

**East access: N34° 4.925', W118° 45.510'     West access: N34° 4.689', W118° 47.624'**

Of all the Backbone segments, this stretch from Corral Canyon Road to Latigo Canyon Road, commonly known as Upper Solstice Canyon or Castro Crest, may have been slammed the hardest by the 2018 Woolsey Fire. With no ocean views, the segment's most notable asset was the longest stretch of riparian woodlands on the entire Backbone Trail. The woods took a real clobbering in the blaze. Some oaks died; others survived with damage. The lovely sycamores lining the canyon creek bed for nearly a mile were pretty much incinerated.

That having been said, this remains one of the most secluded, quiet and pristine parts of the BBT. Although it's far less shady than before, it still offers a rare retreat from "civilization," with virtually no manmade structures to be seen anywhere, except for the prominent cluster of radio towers atop Castro Peak.

The trail actually visits the bottom of two different canyons, Upper Solstice Canyon and Upper Newton Canyon, separated by a gentle mountain pass. Both ends of the hike start out on high chaparral-covered ridges, but the farther you go, the more the canyon slopes close in and create a sequestered hideaway with seasonal creek music, hardy old oaks trees, views of rocky ridges, and a hidden meadow where you might spot deer.

When it's not following the canyon streams, the trail winds across a couple of sheltered upland valleys and up over the saddle, providing sweeping upcountry views of unmarred chaparral-covered slopes. If you come here in the spring, you might be treated to some extraordinary patches of wildflowers and blooming yucca as well.

With their seasonally reliable streams, the canyons' vegetation will slowly recover, barring another fire. Chaparral is already returning. Oaks will continue refoliating and sycamores will resprout. Until then, hikers will have to be content with the peace and solitude that this hike continues to offer.

This trip requires hiking boots to handle loose rocks and gravel along with a plethora of seasonal stream crossings. It can be hiked as a hefty round-trip of 8.2 miles or as a moderate one-way westbound hike, requiring either a second car or a shuttle car and driver. (Due to the remoteness of both trailheads, relying on a rideshare service is not recommended).

**Directions to trailhead (if hiking out-and-back):** From Pacific Coast Highway, head north on Corral Canyon Road. Take it up the mountain for 5½ miles until the pavement turns to gravel, then continue another ¹/₁₀ mile and park in the dirt lot.

**Directions to trailheads (if parking two cars for a one-way hike):** First, park your pick-up car at the end of the hike. From Pacific Coast Highway, head north on Latigo Canyon Road and take it for 7.2 miles into the mountains. Park in the large, dirt parking lot to the right of the road about .4 miles past mile marker 3.66 and before mile marker 3.02. (If you're coming from the 101 freeway, exit at Kanan Road and take it south for 6.6 miles. Half a mile after crossing Mulholland Highway, turn left onto Latigo Canyon Road and continue south for another three

*View of Upper Solstice Canyon and ocean from Castro Ridge (side trip)*

miles. Shortly after passing mile marker 3.02, park in the large, dirt parking lot to the left.) A sign at the trailhead[1] adjoining the lot indicates it's the Backbone Trail.

Next, drive your drop-off car to the start of the hike. Head south on Latigo Canyon Road for 7.4 miles until it ends at the coast. Turn left and take PCH east for 0.7 miles, then turn left on Corral Canyon Road. Take it uphill for 5½ miles until the pavement turns to gravel, then continue ¹/₁₀ mile and park in the dirt lot.

---

*SIDE TRIP: Wish you had some ocean views along this segment? For a side trip that includes a sampling, as well as a fun bit of wild-west-style rocky outcrops, consider an out-and-back trek of about 2¼ miles along Castro Motorway.*

*Follow the dirt road which heads north and uphill from the parking lot. Pass a junction with a dirt road on the left at .3 miles,[2] then another with a dirt road on the right at .4 miles.[3] You'll get good views looking south over the bowl of Upper Solstice Canyon toward the ocean, Palos Verdes and Catalina Island. After the second junction, the road favors the north side of the ridge and offers great views of the San Fernando Valley to the east and Malibu Creek State Park directly below, with Malibou Lake (sic) and Thousand Oaks beyond.*

*At .8 miles, Bulldog Motorway joins the road from the right.[4] Here, a use path leaves from the south side of Castro Motorway, directly opposite Bulldog's terminus. The tiny path leads to the top of a knoll offering a 360° view of the region – the best view on this segment. You can see the entire coastline, the Malibu Creek area, distant northern mountain ranges, Castro Peak and more. (To reach the knoll, follow the gravelly path uphill, then angle to the left up the ridgeline for about two minutes.)*

*Continuing on Castro Motorway, at mile one you'll enter a dramatic outcropping of tan and orange Sespe sandstone. The road squeezes between towering rocks that resemble an old movie setting in which bandits might threaten to jump out. Here a crumbly use trail heads south up to a narrow "hole-in-the-wall" watched over by spooky rock towers honeycombed with cavities – the perfect setting for a Wild West movie shootout.*

*A few minutes later (1.1 miles), where the rocks taper off as you approach a saddle,[5] a tiny path on your left going just few steps off the road will offer you a 180° ocean panorama including the South Bay coastline, Palos Verdes, Catalina Island, tiny Santa Barbara Island and Castro Peak. Down below, a thin thread of riparian woods marks the bottom of Upper Solstice Canyon and the Backbone Trail, which can be seen climbing a pass to the west.*

*This saddle should be your turnaround point. You could continue another ⅓ mile, but at that point a large razor-wire gate (erected by the owner of the communications towers on Castro Peak) blocks further passage.*

**The hike:** The Backbone Trail crosses the parking lot. Locate the starting trailhead[6] on the west side by a sign reading BACKBONE TRAIL, CORRAL CANYON TRAILHEAD. (Be sure to take the single-track trail, not the aforementioned side trip dirt road which climbs the mountain.)

After descending about 100 feet down a single switchback, the trail begins contouring along the northern slope of a basin that drains into Solstice Creek. This highland valley is sheltered on all sides, and supports a variety of chaparral: chamise, monkey flower, dodder, ceanothus and laurel sumac.

For its first ⅔ mile, the trail meanders gently up and down with little real elevation change, offering plenty of views of the open rolling countryside. The sandy trail is often rutted with bicycle tracks.

Half a mile into the hike, the single-track trail joins up with an old gravel road,[7] built to provide access to a powerline tower. Make a gentle right on the road.

Within a minute, the road joins another dirt road at a "T"-shaped junction.[8] (The road to the south ends at a powerline tower; northbound heads steeply up to Castro Motorway – see "Side Trip"). At this junction, the Backbone Trail continues straight, reverting back to a single-track trail and leaving the gravel roads behind.

After the junction, the trail begins a ¾-mile descent down a side canyon toward the main canyon of Solstice Creek.

Where the trail rounds the head of the side canyon and crosses its stream at ¾ miles, a small 20-foot seasonal waterfall lies just upstream to your right. The trail then descends the side canyon more briskly as the route pulls away from the streambed.

The trail continues steadily down the side canyon for another ⅕ mile. After a brief uphill section, a single switchback will take you down a steep, sandy and arid descent to the streambed.

As the trail continues descending, the canyon walls grow higher. After two more stream crossings, the trail heads through a small patch of grass along the brook, then crosses a (burned) live oak woodland.

At 1¼ miles, you'll reach the mouth of the side canyon where it empties into the larger Upper Solstice Canyon. The trail turns right and heads up the main canyon at a barely discernable incline, past coast live oaks and sycamores. This riparian section will last more or less for the next mile, an exceptionally long stretch for the Backbone Trail.

After three more stream crossings, at 1.4 miles the trail skirts the edge of a large oak-ringed meadow, hidden from all signs of civilization, and a good place to spot wildlife late in the day.

A third of a mile and four stream crossings later, the trail crosses through a particularly open stretch for a minute, offering wide views to the right of rock-studded Castro Crest.

After another ¹/₁₀ mile and three more stream crossings, you'll start the 380-foot climb out of the canyon. The trail angles up the south slope through a sprawling patch of coast live oaks. With the stream gone, the trees soon thin out and eventually transition to open chaparral, offering a fine vantage point looking across the canyon toward the stony ridge of Castro Crest as you double back up a single, long switchback.

Ascend the south side of the canyon at a moderate incline. At 2¼ miles, after the trail crosses a metal pipe, its character changes, becoming more gravelly, sandy and steeper as it climbs up the canyon.

About ¹/₁₀ mile later you'll reach the start of a lengthy single switchback, which offers a sweeping vista looking back down the canyon. A long ribbon of oaks and sycamores can be seen lining the bottom, with nothing around but chaparral-covered peaks. In the distance are Saddle Peak and its neighboring summits. If

*The pristine highland basin of Upper Solstice Canyon*

you look closely, you might be able to spot the Castro Crest Parking Lot at the top of Corral Canyon Road where your hike started.

From the switchback onward, the climb becomes downright steep, a true huffer-puffer as you contour up the canyon's south slope on the gravelly and slippery trail. At 2½ miles, it climbs to a "T"-shaped junction[9] atop a large, exposed slab of blue/grey sandstone. Turn right. (The tiny path to the left heads southeast for a few minutes to join the single-track remnants of Borna Drive.)

The Backbone Trail follows the rock slab steeply uphill. It's badly eroded, with plenty of sand-on-rock, making for a "perfect storm" of slipping conditions. Many a mountain biker has lost control here and the drop-off below the trail is quite a steep one.

*A hidden meadow in the middle of Upper Solstice Canyon (photographed before the Woolsey Fire)*

After you leave the slab a few minutes later, the trail turns northward, climbing more gently and offering excellent views looking back down the canyon. It's a pristine setting: gentle mountain slopes with virtually no sign of human encroachment anywhere, except for the trail itself and maybe a far-off house. On a clear day you can see the distant San Gabriel Mountains.

At 2.7 miles, you'll come to a pass between Solstice Canyon and Newton Canyon, comprised of the same blue/grey sandstone. Here at 2,270 feet, you're at the high point on this segment, and you'll get excellent views into both canyons. Just to the north, Castro Peak looks down from its 2,826-foot summit, the highest in the central Santa Monica Mountains. From 1925 to 1971 the top housed a fire lookout tower; now it is loaded with commercial radio towers.

At the pass, you'll meet a gravel road, Newton Motorway,[10] which heads south to Latigo Canyon Road (but climbs over 100 feet and doesn't make for a suitable shortcut to the Backbone Trailhead). The road also heads north to the summit of Castro Peak – but don't bother hiking up there. The top of the peak is private property, and the radio tower tycoon who owns the summit has blocked Newton Motorway partway up the mountain with a large razor-wire gate covered top-to-bottom with signs making it obnoxiously clear that you are not welcome to cross. One reads WHAT PART OF 'NO TRESPASSING' DON'T YOU UNDERSTAND? and another displaying crosshairs is marked IF YOU CAN READ THIS YOU ARE IN RANGE. Okay, dude, we get it. You're a nature-lover.

From the pass, the Backbone Trail follows the gravel road briefly to the right, then within a few steps heads off to the left[11] where a sign reads BACKBONE TRAIL. LATIGO CANYON TRAILHEAD 1.4 MILES.

After leaving the pass, the trail immediately crosses a stand of yucca, brilliant white when blooming in the springtime. It descends very gently along the rim of the upper part of Newton Canyon through a mix of toyon, laurel sumac, chamise and ceanothus.

A few minutes farther down the trail, if you look back toward the pass, on the south side of the canyon you'll see more of that blue/grey scoured rock face, which stands out from the surrounding terrain.

After several minutes of gentle up-and-down, you'll begin a short but moderately steep climb of about two minutes to a sharp left turn in the trail atop the shoulder of a ridge at mile three. This curve offers a fine little lookout point just a few steps off of the trail. From here you can see several miles of Latigo Canyon Road winding through such a number of hairpin curves that it appears to be auditioning for a car commercial. You can also spot the Latigo Trailhead parking lot (your destination), plus Newton Canyon, Zuma Canyon and Buzzard's Roost Peak.

After turning left, the trail descends steeply for several minutes down an arid ridge through low chaparral, mostly monkey flower and dodder. The descent can be gravelly and slippery at times.

About ⅕ mile farther on, the trail lessens its descent as it switchbacks down into the upper reaches of Newton Canyon. As you approach the bottom of the canyon at 3.4 miles, things change dramatically. You'll leave the chaparral zone and enter a short but dense stretch of old oak woods. The steep-walled gorge gives the whole area a tucked-away feeling.

After crossing the stream at 3½ miles, the trail heads up the opposite side of the gorge – a moderate climb of about 170 feet, utilizing a few steps at the beginning.

A few minutes later, a steep little switchback offers a glimpse of the trail right below. Continue climbing moderately past the oaks, cutting across a hillside of exposed crumbly shale. As you near the top, the precipitous drop below the trail reveals a beautiful view looking down into the gorge. Castro Peak looms beyond with its metal towers on top.

You'll reach the top of your climb at mile four. Latigo Canyon Road is visible directly ahead. In spring, this final stretch can be a veritable garden of beautiful orange/yellow monkey flower blooms so profuse that you'd swear someone planted them.

Continue gently downhill to the ending trailhead[1] and dirt parking lot at Latigo Canyon Road, 4.1 miles from where you started.

*Latigo Canyon Road seen from the trail*

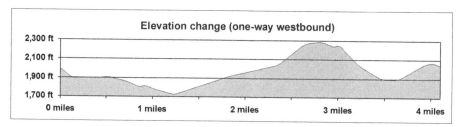

Elevation change (one-way westbound)

| | | | | |
|---|---|---|---|---|
| 2,300 ft | | | | |
| 2,100 ft | | | | |
| 1,900 ft | | | | |
| 1,700 ft | | | | |
| 0 miles | 1 miles | 2 miles | 3 miles | 4 miles |

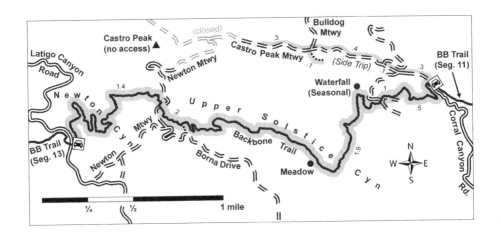

1. N34° 4.689', W118° 47.624'
2. N34° 4.990', W118° 45.716'
3. N34° 5.028', W118° 45.800'
4. N34° 5.109', W118° 46.224'
5. N34° 5.122', W118° 46.474'
6. N34° 4.925', W118° 45.510'
7. N34° 4.867', W118° 45.775'
8. N34° 4.846', W118° 45.806'
9. N34° 4.762', W118° 46.987'
10. N34° 4.850', W118° 47.107'
11. N34° 4.858', W118° 47.102'

## SEGMENT 13:
# LATIGO CANYON ROAD TO KANAN DUME ROAD
### ("NEWTON CANYON TRAIL")

**BBT segment length: 2.2 miles**
**Day hike length: 2.2 miles one-way**
**Suggested day hike: west to east, out-and-back**

**Elevation gain westbound: 240 feet**
**Elevation gain eastbound: 740 feet**
**Difficulty: moderate – Shade factor: 20% (65% pre-fire)**

**Bikes allowed – Horses allowed – Dogs allowed on leash**

**East access: N34° 4.673', W118° 47.640'**     **West access: N34° 4.542', W118° 48.922'**

Although this segment is also referred to as the Newton Canyon Trail, the namesake is a bit misleading because it barely dips down into the canyon, instead spending most of its time on the ridge above. A more accurate choice might have been to name it after the nearby Snakebite Ridge Road, but that sounds pretty horrible, and this hike is anything but. It's a simple, peaceful walk out in the country along serene, oak-dotted hillsides.

This hike doesn't offer much in the way of ocean views, summits or deep forests... but that's not the point of Newton Canyon. The goal here is the journey and not the destination. The canyon's most notable feature is its gentle slopes dotted with old coast live oaks, a surprising number of which managed to survive the catastrophic Woolsey Fire of 2018. Even with limited foliage, the oaks cast their bits of shade your way as you amble along out in the country.

The mountains here, like the hike itself, have a tranquil, bucolic feel to them. They're not very rugged or craggy, but have a more rounded look and are checkered with farmland, giving you the feeling of being in a sheltered rural locale like you might find back east. An ever-present hillside vineyard only adds to the pastoral vibe.

Because most of its incline is fairly tame, this trail is good for those whose condition is just "intermediate." While easier segments of the Backbone Trail exist (such as the Clark Ranch, Fossil Ridge, Dead Horse or Henry Ridge segments), this one will make you feel

*Kanan Dume Road from above the tunnels*

like you took a *real* hike without stressing your system much. The round-trip elevation change of 1,000 feet feels easier than the digits suggest. But do be aware that the toughest parts are at both ends. The western kick-off leaving Kanan Dume Road can seem discouraging as you trudge right up the steep, rocky slope over the twin road tunnels, and the moderately steep eastern end near Latigo

*Coast live oaks line much of the trail*

Canyon Road can be a challenge as well. The rest makes for a pleasant stroll.

**Directions to trailhead:** From Pacific Coast Highway, take Kanan Dume Road north for 4.5 miles. Shortly after emerging from the tunnel, park in the large, paved lot on the west (left) side of the road.

If you're coming from the 101 freeway, exit at Kanan Road (aka Kanan Dume Road) and take it south for 7.9 miles, then park in the large, paved lot on the west (right) side of the road.

The trailhead is one of the most popular and most developed of any on the Backbone Trail. It includes a bench under a little shelter, an information kiosk, well-maintained pit toilets, trash bins, plenty of signage and maps. There is, however, no drinking water here. When the lot fills up on busy weekends, use the unpaved overflow parking lot 1/3 mile north on the west side of the road.

**The hike:** From the parking lot, two segments of the Backbone Trail head off in different directions. Don't take the trail that heads past the bathroom; instead follow the trail to the left that leaves the parking lot heading south, closely paralleling Kanan Dume Road uphill and toward the tunnels. A sign at the trailhead[1] reads BACKBONE TRAIL; KANAN ROAD TRAILHEAD; LATIGO CANYON TRAIL 2.5 MILES. (The sign's mileage is slightly off.)

The trail starts off in no-nonsense mode, trudging directly up a steep, rocky and sun-exposed dirt roadbed right above noisy Kanan Dume Road – hardly a back-to-nature experience. Take heart, as this brief stretch is by far the *worst* part of the hike.

At the top of the climb at .2 miles, a tiny side path[2] heads off to the right.

*SIDE TRIP: A detour of just a few steps down this path will take you to a viewpoint looking south down Ramirez Canyon from the top of the road tunnels, with the ocean in the distance. (The tentative little path continues south along the top of the bluffs high above Kanan Dume Road, but travel on it is not recommended as it dead-ends in 1/5 mile at a disappointing viewpoint above the road.)*

From the little junction, the Backbone Trail turns left and follows the old roadbed over the tunnels. From here onward, your hike will become more pleasant as the well-constructed trail (built by the Youth Conservation Corps in 1986 as one of the first sections created solely for the Backbone Trail) heads east into more hospitable environs.

*The Rosenthal Vineyards carpet a hillside far below the trail*

Within a few more minutes, the trail curves away from the tunnels and continues behind the north side of the ridge. Things change for the better as you enter the first of several hillside oak tracts that will become the defining characteristic of much of this hike. You'll enjoy views looking back into Zuma Canyon and ahead into Newton Canyon with its sprawling hillside vineyards. Diminishing noise from Kanan Dume Road will be audible for the first third of the hike.

At .4 miles, the trail pitches downward and angles left, while the overgrown remnants of Snakebite Ridge Road branch off to the right.[3]

1/10 mile farther, the trail crosses a paved private road,[4] which heads up to a ridgetop vineyard and private property, eventually joining Snakebite Ridge Road. Cross the road and continue straight on the single-track trail.

For the next half mile, the trail climbs the ridge at a moderate pace, passing dozens of old oaks. A short break from the oaks comes at .7 miles when you'll briefly cross a sage-covered hillside, offering a decent view looking north at Kanan Dume Road as it meanders through the mountains.

Within a few minutes, you'll leave the oaks a second time as the trail curves its way up across another longer stretch of sage-covered hillside. The clearing offers sweeping views out over Newton Canyon, including your final views of Kanan Dume Road. Looking east from the center of the clearing, you'll see Castro Peak, the flat-topped mountain crowned with radio towers, standing well above its neighbors. At 2,826 feet, Castro is the high point in the central Santa Monica Mountains.

The quilt work patches of agricultural land that you see 400 feet below are the Rosenthal Vineyards, growing Cabernet Sauvignon and Chardonnay. If you're curious about the wines that come from these grapes, Rosenthal operates a public tasting room on Pacific Coast Highway.

In a few minutes, you'll find yourself back under more oaks, and at .9 miles, the trail tops a local high point of about 1,900 feet slightly below the crest of the ridge. Just a few steps before the high point, a side path heads up to the right[5].

*SIDE TRIP: For those of you hungering for a little ocean view on this hike, a brief trip up the side path will give you a vista of the Pacific.*

*After turning onto the path, immediately angle to the right up the hill. In about half a minute, you'll intersect with the overgrown remnants of Snakebite Ridge Road[6] atop the ridge in a meadow which can be gorgeous in the spring with purple lupine and non-native yellow mustard. Follow the roadbed westbound as it curves to the left, out onto a flat-topped grassy promontory about 1/10 mile from where you left the Backbone Trail.*

*From the promontory you'll enjoy a panoramic view of the area: the Pacific sprawls to the south beyond a ridge across Ramirez Canyon and Kanan Dume Road. Catalina Island and Palos Verdes are visible to the south-southeast, while pointy little Santa Barbara Island is to the south-southwest, and farther to the right, San Nicholas Island may be visible over 60 miles away. Turning some more to the right, you'll see Buzzard's Roost Peak to the west, followed by Boney Mountain and the pinnacles of Saddle Rock. Completing the circle to the right, the vineyards in Newton Canyon are visible to the north, followed by Castro Peak to the northeast.*

*The road continues downhill but enters private property. Turn around here.[7]*

The Backbone Trail hugs the ridgeline for a few minutes, then begins a gentle descent of about 170 feet into a tributary of Newton Canyon, alternating between tracts of oaks, patches of reeds, and chaparral. You'll continue to get views to the north looking over the canyon, with Latigo Canyon Road twisting its way high on the opposite side, the vineyard down below, and Castro Peak beyond.

At 1.4 miles, when the trail finally reaches the tributary canyon bottom, you'll cross the first of two small streams. Here the oaks cluster thicker, a bit of moss lines the trail, and the sound of birds often fills the air.

Cross the seasonal stream, then zigzag uphill. You'll soon crest a low divide between the two streams and within a minute descend to the second stream, crossing it at 1.71 miles. This second mini-canyon is more moist than the first, but you will soon leave it behind as the trail turns westbound, working its way up the ridge and back into chaparral.

About five minutes past the crossing, the trail makes a sharp turn back to the east and begins its final ascent towards Latigo Canyon Road. The climb is only moderately steep but relentless as it heads up the arid south-facing side of the canyon through sunbaked chaparral.

*A quick peek at the Pacific from the top of Snakebite Ridge (side trip)*

You'll cross under some powerlines at mile two, signaling the home stretch of the hike. The occasional sound of Latigo Canyon Road is discernable just uphill to your left.

At 2.1 miles, the trail rounds the head of the tributary canyon via a sharp right curve, then makes its final climb to Latigo Canyon Road through a patch of chamise. In the spring and early summer, this open hillside is full of wildflowers. From here you'll get one of the best views on the hike, looking back down the length of Newton Canyon. You can see the Rosenthal Vineyards, Kanan Dume Road, and beyond it, Zuma Canyon. On a clear day you can also see the outline of Boney Mountain in the distance.

The trail reaches Latigo Canyon Road[8] and the end of this segment at 2.2 miles. A perfectly smooth, bun-shaped boulder here makes an excellent sitting spot under the shade of an oak tree. The gravel parking lot directly across the road marks the start of the next segment. From here you can continue on to Segment 12 or return the way you came, an easier return trip which will be mostly downhill.

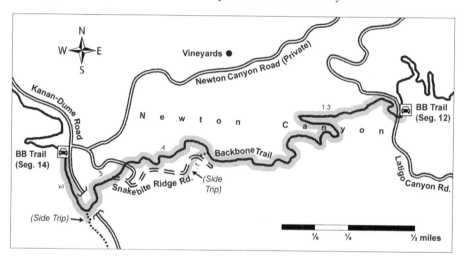

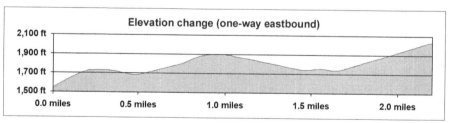

GPS COORDINATES

1. N34° 4.542', W118° 48.922'    2. N34° 4.373', W118° 48.839'    3. N34° 4.469', W118° 48.766'
4. N34° 4.515', W118° 48.703'    5. N34° 4.553', W118° 48.371'    6. N34° 4.554', W118° 48.397'
7. N34° 4.509', W118° 48.470'    8. N34° 4.673', W118° 47.640'

# KANAN DUME ROAD TO ZUMA RIDGE MOTORWAY
## ("UPPER ZUMA CANYON TRAIL")

**BBT segment length: 2.5 miles**
**Day hike length: 2.5 miles one-way**
**Suggested day hike: east to west, out-and-back**

**Elevation gain westbound: 561 feet**
**Elevation gain eastbound: 374 feet**
**Difficulty: moderate – Shade factor: 20% (50% pre-fire)**

**Bikes allowed – Horses allowed – Dogs allowed on leash**

**East access: N34° 4.560', W118° 48.934'     West access: N34° 5.440', W118° 50.308'**

"Zuma" is the Chumash word for "abundance," an apt name for both the canyon and this segment. In spite of the 2018 Woolsey Fire, even now this place still offers a little bit of everything. This moderate hike takes you down into beautiful, deep Zuma Canyon, with its oak woods and frequently running stream. A dramatic stretch across a steep grassy bluff offers views looking south through rugged Lower Zuma Canyon, with its crags and cliffs, to the Pacific Ocean beyond.

But soon you'll find yourself on a hike of more subtle beauty as the trail enters Upper Zuma Canyon, a protected highland valley surrounded by miles of gentle brush-covered mountains, where the chaparral closes in to create a cozy feeling. Up here, you might even come upon some surprising patches of tall reeds.

You won't see much ocean on this hike, but you will see something we don't get much of on the Backbone Trail: waterfalls. Two of them, in fact. Upper Zuma Falls is an impressive sight after rains but is often dry otherwise, while cute little Newton Canyon Falls, tucked away in its rocky nook, usually has some water in it.

This segment is relatively easy on the feet, save for a few rocky patches, making for a good hike if you want to get away from it all without busting your hump too much.

**Directions to trailhead:** From Pacific Coast Highway, take Kanan Dume Road north for 5.1 miles. Just after the tunnel, park in the large, paved lot on the west (left) side of the road. If you're coming from the

*At the trailhead*

*Rugged Lower Zuma Canyon, seen from the Backbone Trail, leads south to the ocean*

101 freeway, exit at Kanan Road (aka Kanan Dume Road) and take it south over the mountains for 7.9 miles, passing through two tunnels. Just before the third tunnel, park in the large, paved lot on the right (west) side of the road.

The trailhead is one of the most popular and most developed of any on the Backbone Trail. It includes a bench under a little shelter, an info kiosk, pit toilets and trash bins, but no drinking water. When the lot fills up on busy weekends, use the unpaved overflow parking lot ⅓ mile north on the west side of the road.

**The hike:** From the parking lot, two segments of the Backbone Trail head in different directions. Take the trail that heads west past the bathrooms.[1] (Avoid the other trail, which leaves the parking lot heading south and parallels Kanan Dume Road uphill towards the tunnels.) A few seconds into the walk, a sign reads BACKBONE TRAIL, KANAN ROAD TRAILHEAD. ZUMA RIDGE MOTORWAY 2.5 MILES.

Within a minute, you'll begin to descend the south slope of Newton Canyon (an offshoot of Zuma Canyon) at a moderate pace. A few minutes into the hike, a sharp right switchback offers an excellent view of the big bend in Zuma Canyon, looking both up and downstream. Far off at the head of the canyon, the shadowy cleft of Upper Zuma Falls is discernable on a large, tan-colored cliff.

At 0.3 miles, after a few minutes of steeper descent, you'll reach the bottom of Newton Canyon, where you'll cross an offshoot of Newton Stream in a brief patch of riparian woodland.

*SIDE TRIP: About half a minute past the stream, a small side path[2] to the left will take you to Newton Canyon Falls, a cute little waterfall of about 25 feet that often contains water, set in a rocky nook. To reach the brink of the falls, turn left off of the BBT onto the little path. Then take a second left[3] after about 15 seconds, heading down the even smaller path. In another 15 seconds or so you'll arrive at the brink.*

To access the bottom of the falls (a round-trip of less than .2 miles), continue straight at the last junction, instead of taking the left path to the brink. In about 1½ minutes, at another small junction,[4] follow the path to the left steeply downhill into the gorge. (The other fork, straight and uphill, leads nowhere and peters out.) Less than half a minute later, take another left at a minor T-junction.[5] Follow the steep use path down to the little hideaway at the base of the falls, where the water slides down a twin-chuted rock face of beige sandstone into a moss-covered, cliff-ringed basin.

After crossing the stream, you'll begin heading up the sunnier north slope of the canyon through a dry chaparral mix of toyon, laurel sumac, ceanothus and sage. The trail surface becomes crumbly and rocky, making for some hard walking. Noise from Kanan Dume Road becomes more prevalent as you climb closer to the road above. About half a mile into the hike, you'll reach a local high point on the trail where a spur path[6] heads off sharply to the right.

*ALTERNATE ROUTE: If on your return you're wiped out and would like to shorten your route as well as avoid the climb in and out of Newton Canyon, you can take this short path to the overflow parking lot[7] on Kanan Dume Road, then walk back to your car along the shoulder of the road. This will shorten your trip by 0.2 miles and 120 feet of elevation gain. The shoulder is more than wide enough for pedestrians but walking along busy Kanan Dume Road can be less than pleasant.*

From the junction, the Backbone Trail continues left through an impressive stand of yucca. At times their decayed, splintered fronds can be seen littering the trail, resembling tiny old-fashioned straw brooms.

A few minutes later, the trail traverses a steep grassy hillside, offering impressive views looking south down Lower Zuma Canyon, with its rugged crags and cliffs, toward the Pacific Ocean beyond. On a clear day or in patchy fog, this can be one of the more memorable views along the entire Backbone Trail. It's also the only glimpse you'll get of the Pacific on this segment.

You'll soon leave the grassy hillside and begin a moderate ⅓-mile descent to the bottom of Zuma Canyon through sage and laurel sumac. As you near the bottom, the pace eases up and the noise from Kanan Dume Road is replaced by the much more melodious sound of Zuma Creek gurgling below.

*The footbridge over Zuma Creek is slated to be rebuilt*

The trail reaches the bottom of Zuma Canyon at mile one, crossing the creek in a woodland of old oaks. The sturdy footbridge which crossed the creek here was destroyed in the Woolsey Fire of 2018. As of this publication (2021) the bridge is still missing, but plans are in the works to replace it.

Immediately after the crossing, the trail begins a moderate climb up the south side of the canyon, first past some old overarching oak trees, then through chaparral.

It takes about five minutes to reach the top of the climb. Here you'll get your first views of the upper canyon, including Saddle Rock, a prominent landmark to the north of Mulholland Highway, which you will continue to see for much of the hike. It looks like

*Old oaks arch over the trail at the canyon bottom*

the ruins of an ancient twin-towered castle atop a mountain. From the crest, you'll descend gently through chaparral past the occasional oak tree.

At 1⅓ miles, you'll cross a side stream that is often dry, although after major rains you may get your feet a little wet. These side streams can provide a surprising respite from the heat in the late afternoon or early evening as cool air slides down their little canyons, acting as a natural air conditioner.

The fully sun-exposed trail continues uphill to top a small rise at 1.6 miles. A false trail crosses the main trail here[8] but quickly dies out in both directions.

For the next ⅓ mile, the Backbone Trail undulates up and down, gradually gaining elevation and crossing several minor streams. With every step, the upper canyon becomes gentler and more shallow, resembling a wide valley more than a canyon. You might hear the croaking of frogs or chirping of crickets, as well as the occasional motorcycle zooming along Kanan Dume Road far away. It makes for easy hiking with plenty of time for the mind to wander.

At about 1.9 miles, you'll see Upper Zuma Falls straight ahead across the canyon. (This spot is sometimes misidentified on websites as Newton Canyon Falls, but you're far from Newton Canyon at this point.) It's often dry, in which case it appears as a dark gash in a tan cliff with some white stains around it, but after a rain it can be quite impressive.

About a minute and a half later, when the trail makes a sharp left turn, you'll get your best (and last) view of the falls across the canyon. The trail promptly zigzags to cross a seasonal stream. This area is marshy enough to support a thicket of reeds, an unusual sight for the Santa Monica Mountains.

After crossing the stream, the trail makes its second sharp left in the zigzag. Here, at mile two, a small side path leads off to the right.[9]

*SIDE TRIP:* If you'd like a closer look at Upper Zuma Falls, a short detour on this overgrown little path of roughly ¹⁄₁₀ mile will take you to a clearing where you can get a better vantage. The path heads downhill for about 30 seconds, then turns left and follows Zuma Creek through grass and thickets, becoming fainter as you proceed. After about another minute, the path crosses the creek, then soon fades away into a small clearing that offers a good view looking up at the falls. (Although it is technically possible to bushwhack from here to the base of the falls, the route is virtually impassable and further travel on it is not recommended.)

From the zigzag, the Backbone Trail begins winding its way steadily up a side canyon, in and out of thickets of chaparral (chamise, laurel sumac, yucca, sage, ceanothus and western mountain mahogany).

As you ascend the side canyon, progressively better views open up looking back down the length of Upper Zuma Canyon. High on the mountainside, the rock cut of Kanan Dume Road is visible, as well as a distant house or two. No other signs of human intervention mar the miles of chaparral-covered slopes.

At 2¼ miles, as you continue climbing, you'll cross a streambed with another thicket of reeds and a little marsh. Just beyond, the trail enters a lovely forest of old coast live oaks and climbs up its slope at a steady moderate pace.

¹⁄₁₀ mile later, the climb up the side canyon diminishes to a gentle pace as you emerge from the oaks back into ceanothus-dominated chaparral.

At 2½ miles, the single-track Backbone Trail intersects with a dirt road, Zuma Ridge Motorway, at a "T"-shaped junction.[10] Seemingly out of nowhere, the sound of traffic along Encinal Canyon Road becomes audible and the wilderness sensation of Upper Zuma Canyon slips away.

This is the end of the segment. A sign here reads ZUMA RIDGE TRAIL; ENCINAL CANYON ROAD .3 MILES; BUSCH DRIVE 5.4 MILES. From here you can return the way you came or continue on to Segment 15 of the Backbone Trail, which follows the dirt road to the left and uphill. (Or take the use path that heads left and reconnects with the Backbone Trail in ¹⁄₁₀ mile.)

If you care to make this a "destination" hike, you can always continue on the side route up to the top of Buzzard's Roost for wonderful views – definitely a "destination" (see side trip in Segment 15 chapter).

*Upper Zuma Falls after a rainy period*

*ALTERNATE EXIT: If you're hiking one-way or wish to exit the backcountry, from here you can take Zuma Ridge Motorway to Encinal Canyon Road in under ten minutes of easy walking. To do so, turn right on the fully sun-exposed dirt road and follow it over a rise about .2 miles from where you left the BBT. The road heads downhill through a gate and past a small water recycling plant before ending at Encinal Canyon Road in .4 miles. Parking is available near the terminus[11] in a large, dirt turnout on the south side of Encinal Canyon Road just west of mile marker 0.57 and east of mile marker 0.85.*

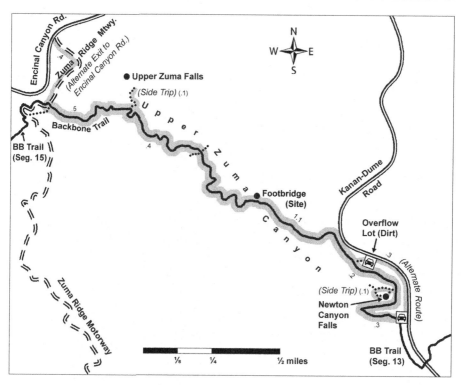

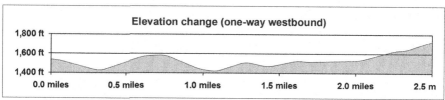

GPS COORDINATES

1. N34° 4.560', W118° 48.934'   2. N34° 4.634', W118° 48.944'   3. N34° 4.635', W118° 48.953'
4. N34° 4.648', W118° 49.000'   5. N34° 4.644', W118° 48.997'   6. N34° 4.729', W118° 49.114'
7. N34° 4.724', W118° 49.083'   8. N34° 5.069', W118° 49.725'   9. N34° 5.211', W118° 50.004'
10. N34° 5.210', W118° 50.353'   11. N34° 5.440', W118° 50.308'

# SEGMENT 15:
# ZUMA RIDGE MOTORWAY TO ENCINAL CANYON ROAD
## ("TRANCAS CANYON")

**BBT segment length: 2.4 miles**
**Day hike length: 2.4 miles one-way**
**Suggested day hike: west to east, out-and-back**

**Elevation gain westbound: 360 feet**
**Elevation gain eastbound: 708 feet**
**Difficulty: moderate – Shade factor: 10% (20% pre-fire)**

**Bikes allowed – Horses allowed – Dogs allowed on leash**

**East access: N34° 5.444', W118° 50.309'     West access: N34° 5.124', W118° 51.749'**

This segment takes you down into quiet Trancas Canyon. Trancas is not one of the more well-known canyons, and the trail running through it is lightly visited because until recently the easternmost section of it crossed private property. In fact, you just might be the only people venturing into this place during your visit.

Trancas Canyon has a simple, peaceful charm to it. The wide and flat canyon bottom shelters an expansive forest of ancient oaks and other riparian flora, one of the most charming surprises to be found along the entire Backbone Trail. This fact is something of a miracle, considering that in 2018, Trancas had the misfortune of being nearly the geographic center of the colossal Woolsey Megafire, which

*The trail through the old oak forest*

raged through the canyon, burning its forest, destroying its pair of picturesque footbridges and devastating nearly everything for several miles in all directions. Somehow, the majority of these tough old trees managed to survive and are gradually regrowing their once thick shade canopy.

Several old footpaths and abandoned roads crisscross the canyon, slowly returning back to nature as their overgrown routes become carpeted with flowers, grass and saplings – a reminder that what was once developed can return to its original state, given time.

The majority of this hike traverses the pristine, chaparral-covered ridges overlooking the canyon, offering some fine views over the sprawling woods below and down the canyon to the ocean. After leaving Trancas, the trail gently winds its way up through open backcountry along the rim of a small upland valley, crossing the divide between Trancas and Zuma Canyons. At times it can be virtually silent up here, with relatively few hikers or mountain bikers venturing into this parcel that until 2016 was private land co-owned by Arnold Schwarzenegger.

Trancas is a good area for spotting wildlife. One morning, while checking out the eastern stretch high above the canyon, one of your authors (Doug) spotted a mountain lion for a few thrilling seconds – our only sighting in these mountains. We've also seen fresh deer tracks in the mud along the trail (see photo page 43).

All in all, this hike will leave you feeling like you've gotten away

*Waves break on Zuma Beach 2,450 feet below Buzzard's Roost*

from it all without an inordinate amount of effort. The elevation change feels easier than the numbers suggest, thanks to the well-graded and smartly designed trail.

**Directions to trailhead:** From Pacific Coast Highway, take Encinal Canyon Road north for nine miles (be sure to follow the road when it turns right instead of continuing straight on Lechusa Road). After mile marker 2.45, turn left onto dirt Clark Ranch Road (directly across Encinal Canyon Road from County Fire Camp 13). Just a few seconds up the road, park in the large, dirt parking lot on the right.

If you're coming from the 101 freeway, exit at Kanan Dume Road (signed Kanan Road) and take it south for 6.2 miles, then turn right onto Mulholland Highway. In another mile, bear left onto Encinal Canyon Road. Continue for 2½ miles, then turn right onto dirt Clark Ranch Road (directly across from County Fire Camp 13). Just a few seconds up the road, park in the large dirt lot on the right.

The Backbone Trail crosses the parking lot. This particular segment leaves the lot at its southeast corner[1] (to your left when facing Encinal Canyon Road). Avoid the temptation to walk back out to Encinal Canyon Road via Clark Ranch Road. To reach the Backbone Trail that way, you must then walk east on Encinal Canyon Road, which can get harrowing when the occasional speed demon whizzes by.

**The hike:** Leave the parking lot and follow the dirt road to the southeast, up over a little rise. In a couple of minutes you'll cross Encinal Canyon Road.[2]

The trail continues across the highway as a dirt road, but within another minute, after passing through an old gate, it narrows and becomes more trail-like. It heads downhill through plenty of chaparral (chamise, ceanothus, laurel sumac, western mountain mahogany, sage and monkey flower) but without significant shade.

.2 miles into the hike, a worn-looking path intersects the trail.[3] (The path to the left works its way up the ridge, rejoining the Backbone Trail in a mile, but a few unnervingly steep sections prevent us from recommending it as an alternate route. The path to the right branches into smaller paths that head onto the private property of Fire Camp 13.) As you continue straight, the last vestiges of the old road soon fade away and the gravelly trail, now single-track, works its way moderately down into Trancas Canyon.

At another minor junction[4] (.4 miles), a gravelly footpath heads to the right and up the ridge, eventually joining the path that you crossed earlier. Steep and rutted, it doesn't make a suitable hike to anything. Continue straight and downhill.

In another $\frac{1}{10}$ mile, you'll reach the canyon bottom just above Trancas Stream. Here, the trail takes a hard left,[5] detouring away from the old route which once continued straight onto a sturdy equestrian footbridge and crossed the canyon high above the stream. Unfortunately, the bridge burned, melted and collapsed in the Woolsey Fire and there are no immediate plans to replace it.

The trail briefly doubles back to a foot crossing of Trancas Stream, which frequently has water in it. Just past the foot crossing, followed by a sharp right turn, the remnants of an old road head off to the left,[6] steeply climbing the canyon. Continue straight on the Backbone Trail, once again heading south.

The trail descends further along a shelf above the seasonally babbling brook, in-and-out of sun-exposed chaparral, including a large stand of yucca.

About .6 miles into the hike, the trail drops down into a sprawling tract of over-arching old oaks. With the stream wandering nearby, these woods offer your best opportunity on this hike for a pleasant time-out. It's beautiful stretches like this which make up some of the very best parts of the Backbone Trail.

At $\frac{2}{3}$ mile, while still under the oaks, the trail turns sharply left at a barely discernable T-junction,[7] veering away from the main stream and heading gently up a tributary canyon. (The tiny path to the right crosses Trancas Stream and follows it down-canyon along the bed of an old road, abandoned decades ago, which once went all the way to the sea. The route, now overgrown with grass and saplings, is reverting back to nature.)

The Backbone Trail follows the tributary away from the main stream, heading up a side canyon through the woods. At .8 miles it makes a sharp left to ford the tributary at the site of a smaller equestrian footbridge also destroyed in the Woolsey Fire, crossing just to the left of the remaining bridge abutment.

*View from the trail looking north over the defunct Malibu Golf Club*

(Just before the crossing, another faint and disused trail heads off to the right,[8] following an old roadbed up the shoulder of the ridge and contouring south along the east rim of Trancas Canyon. While it does technically go through to Trancas Canyon Road, the footpath of 1.1 miles is heavily overgrown in spots, barely passable and not recommended.)

After the crossing, the Backbone Trail angles moderately up the canyon slope through chaparral, steadily gaining altitude above the canyon floor.

At mile one, a sharp right switchback up the mountain offers an excellent vista looking both up and down Trancas Canyon. To the south, you'll get a bird's eye view down on one of the largest expanses of woods visible from anywhere on the BBT. Looking up-canyon, you can see the stretch of

*Saddle Rock seen from Zuma Ridge Motorway*

trail you've just walked, and looking directly across, carved into the opposite side of the canyon, the remnants of an old road that parallels the trail but never connects.

The trail maintains a moderate incline as it climbs the sunny ridge high above the side canyon, through arid chaparral including plenty of yucca. The higher you climb, the more views open up of the side canyon below. You'll also get occasional glimpses of the ocean down past the mouth of Trancas. Eventually a lesser ridge rises up on the right, blocking any further views of Trancas Canyon.

At 1.4 miles, a small side trail[9] leaves the Backbone Trail, heading up to the left. (This is the terminus of the path that you crossed early into the hike.)

**SIDE TRIP:** *A trip up the path will take you to the top of a knoll offering an impressive panorama looking north over the Malibu upcountry. To reach it, follow the steep path up the slope for three minutes until it tops the knoll at a sharp right curve.[10] (The path continues down, rejoining the Backbone Trail in .9 miles, but a few ridiculously steep sections prevent us from recommending it. Best to turn around here.)*

*Looking west from the summit, you can see the Backbone Trail and its intersecting paths, including the old road across Trancas Canyon. Farther to the west is the rugged ridge of Boney Mountain, with Sandstone Peak rising above its neighbors. Turning north, you'll see the Malibu Golf Club just past Encinal Canyon Road, which cuts its way across the landscape. Behind and farther east are the crags along Mulholland Highway, with the twin-spired Saddle Rock at the far eastern end. To the east lies the high but gentle Zuma Ridge, with Buzzard's Roost at its apex. Zuma Ridge Motorway, where this Backbone Trail segment terminates, is visible cutting its way along the spine.*

Once past the junction, the trail lessens its climb as it veers away from the edge of the ridge, gently meandering across an undulating plateau. In a few minutes you'll top a double crest, from which you can see what remains of the closed Malibu Golf Club down to the left. The trail then winds its way uphill at a gentle pace for a few minutes until, at 1¾ miles, it dumps you out at a T-junction[11] on Trancas Canyon Road (a dirt road built to provide access to powerline towers).

*ALTERNATE EXIT: If you want to exit the hike here to Encinal Canyon Road, go left on Trancas Canyon Road, then follow it to the right as it curves downhill. It heads steadily down to Encinal Canyon Road in .3 miles, passing a large gate just before its terminus[12] at mile marker 1.32. A few nearby turnouts allow room for three or more cars.*

From the junction, the Backbone Trail continues to the right on the dirt road. After about two minutes, where the road makes a hairpin right turn at 1.8 miles, the BBT leaves the road and heads straight onto a single-track footpath.[13] (Continuing down the road would take you eventually to a dead-end at a powerline tower.)

After leaving the road, the Backbone Trail climbs steadily along the side of a little highland valley through chaparral and grassland. A plaque on the right at 1.9 miles recognizes the donation of this parcel of land from former governor Arnold Schwarzenegger and Betty Weider, officially completing the Backbone Trail in in 2016.

Roughly 2.1 miles into the hike, the tiny stream catches up to the trail, supporting a bit more vegetation. Within a few more minutes, the trail crosses a grassy meadow near the head of the little highland valley, then turns left and leaves the valley, climbing toward the crest of Zuma Ridge.

At 2.2 miles, make a sharp right at a "T"-shaped junction.[14] (The path to the left quickly ends at a clearing.) Continue up a short, steep scramble to the top of Zuma Ridge. You'll reach the crest in a minute, then start a gentle descent towards dirt Zuma Ridge Motorway.

At about 2¼ miles, continue past a lightly used side path[15] heading steeply up the ridge to the right (eventually connecting to Zuma Ridge Motorway but useless as a shortcut). You'll see the motorway ahead and Encinal Canyon Road just down the hill to the left. Soon after, the trail curves left and sharply downhill, keeping just above a bluff overlooking the motorway to your right. Use caution coming down the steep, gravelly trail.

The single-track trail meets the motorway at 2.3 miles.[16] The Backbone Trail turns left, following the motorway downhill.

*ALTERNATE ROUTE: If you wish to avoid dirt road walking, the use path that continues eastbound directly across the motorway will take you to the end of this segment, bypassing the motorway entirely. After ¹⁄₁₀ mile of steady downhill along the sun-exposed ridge through knee-high scrub and grass, the path rejoins Zuma Ridge Motorway at a junction[17] where the single-track Backbone Trail heads off to the right onto Segment 14.*

*SIDE TRIP: For those of you wishing this segment had a "destination" ending, you can continue on the lengthy side trip up Zuma Ridge Motorway to 2,450-foot Buzzard's Roost. With a distance of 1.3 miles each way and an elevation gain of over 650 feet, this is the toughest of any of our side trips, but the views from the roost justify it.*

*The road is popular with mountain bikers who like to zoom down its steep descent. It's not as well-suited for hikers due to its relentless climb and full sun exposure. On a hot sunny day, it's what some hikers refer to as a "death march," a flop-sweat-inducing, thigh-bursting uphill stomp under nature's giant heat lamp... Fun! It's best attempted on a cool or cloudy morning, or in the evening before sunset.*

*To hike to the roost, turn right on Zuma Ridge Motorway and follow it uphill. The wide dirt road maintains a steady, moderate-to-steep climb up the ridgeline, keeping close to the crest. It occasionally switches sides, providing alternating views east over Zuma Canyon and Saddle Rock and west over Boney Mountain and Sandstone Peak.*

*⅓ mile into the side trip, as the road switches to the west side of the mountain, a small path[16] heads steeply back down the ridgeline to rejoin the Backbone Trail at a point mentioned earlier. Using it as a "shortcut" on your return is not recommended.*

*After several more minutes of climbing, you'll approach a sharp knoll on the right, the high point on Zuma Ridge and your destination for this side trip. As the road turns sharp left at 1.1 miles, a little side path to the right[17] winds steeply up the knoll. Don't bother with this path – a much easier route to the top lies just ahead.*

*At 1.2 miles, just a few steps past the gated entrance to private Buzzard's Roost Ranch, Zuma Ridge Motorway tops a crest. Turn right here onto a smaller dirt road[18] that winds the last tenth of a mile up to the top of the little knoll.*

*At the sandy, sun-exposed summit, a little stone labyrinth and an unforgettable view await. Looking south you can see the long strand of white waves crashing on Zuma Beach. To the southeast, Point Dume juts into the sea. To the southwest yawns 1,500-foot deep Trancas Canyon. To the northwest, Boney Mountain forms a clear landmark, as does Saddle Rock due north. In the distance lie the high ridges of the Sespe Condor Refuge.*

*Turn around here – it's all downhill back to the Backbone Trail.*

At 2.4 miles, after a few minutes of walking eastward on the dirt road Backbone Trail, you'll come to the end of Segment 15 at a rather anticlimactic turnaround spot in the middle of nowhere. Here, where the road bends to the left at a gentle saddle, the single-track Backbone Trail heads off to the right[17] onto Segment 14, down into Zuma Canyon and eventually to Kanan Dume Road. A sign here reads BACKBONE TRAIL, KANAN TRAILHEAD 2.5 MILES.

*ALTERNATE EXIT: If you're hiking one-way or wish to exit the backcountry, you can take Zuma Ridge Motorway to Encinal Canyon Road in less than ten minutes of easy walking. Simply follow the sun-exposed dirt road as it bends to the left, topping a rise at .2 miles, then heading downhill through a gate past a small water treatment plant before ending at Encinal Canyon Road in .4 miles.[21]*

*Parking is available near the terminus in a large dirt turnout on the south side of Encinal Canyon Road, just west of mile marker 0.57 and well east of mile marker 0.85.*

If you don't wish to exit to Encinal Canyon Road, you can either turn around and return back through Trancas Canyon or continue on to Segment 14 through Zuma Canyon.

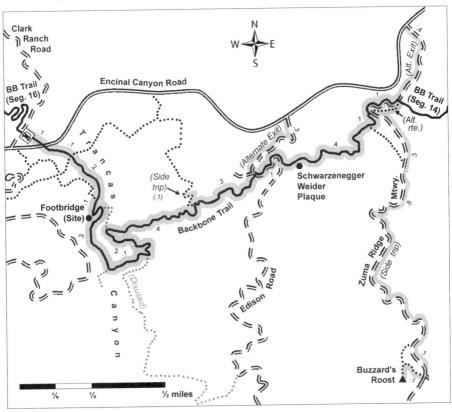

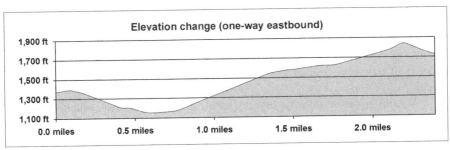

## GPS COORDINATES

1.  N34° 5.124', W118° 51.749'
2.  N34° 5.106', W118° 51.685'
3.  N34° 5.072', W118° 51.588'
4.  N34° 4.952', W118° 51.495'
5.  N34° 4.892', W118° 51.509'
6.  N34° 4.909', W118° 51.488'
7.  N34° 4.751', W118° 51.450'
8.  N34° 4.782', W118° 51.312'
9.  N34° 4.924', W118° 51.145'
10. N34° 4.941', W118° 51.179'
11. N34° 5.005', W118° 50.888'
12. N34° 5.161', W118° 50.787'
13. N34° 5.053', W118° 50.822'
14. N34° 5.113', W118° 50.508'
15. N34° 5.151', W118° 50.454'
16. N34° 5.187', W118° 50.452'
17. N34° 5.210', W118° 50.353'
18. N34° 5.016', W118° 50.333'
19. N34° 4.510', W118° 50.265'
20. N34° 4.437', W118° 50.161'
21. N34° 5.444', W118° 50.309'

# ENCINAL CANYON ROAD TO MULHOLLAND HIGHWAY
## ("CLARK RANCH TRAIL")

**BBT segment length: 1.1 miles**
**Day hike length: 1.1 miles one-way**
**Suggested day hike: east to west, out-and-back**

**Elevation gain westbound: 264 feet**
**Elevation gain eastbound: 22 feet**
**Difficulty: easy – Shade factor: 5% (10% pre-fire)**

**Bikes allowed – Horses allowed – Dogs allowed on leash**

**East access: N34° 5.140', W118° 51.768'      West access: N34° 5.495', W118° 52.030'**

Welcome to one of the easiest segments of the Backbone Trail, offering nothing in the way of drama, but plenty of quiet, serene single-track in the Santa Monica Mountains backcountry. This section is a relatively new addition, constructed with bikes and horses in mind as well as hikers, and the incline is so steady and gentle that you'll hardly notice you're gaining elevation. It makes for a pleasant stroll of less than a half hour one-way, and it's smooth enough that you could possibly do it in regular walking shoes. This is, however, NOT a trail for a hot sunny day. With a shade factor of 5%, it's the sunniest segment of the Backbone Trail.

The trail climbs gently up a sun-exposed escarpment, then heads along the side of a canyon, winding through chaparral, with views down into the canyon and its thicker vegetation below. It's on the return trip, however, when you notice the real views. Looking down towards Encinal Canyon Road, the surrounding countryside and the parking lot below, you'll realize that you've climbed more than you thought, and it makes the hike seem more impressive.

All in all, this is pretty tame terrain by Santa Monica Mountains standards: no sheer drop-offs, no ocean views, just rolling upcountry and chaparral. In spring, the trail hosts a true cornucopia of different wildflowers. In a good year, you'll see blooms of pale blue, yellow, orange, dark purple and peach, as well as white yucca.

A number of small footpaths crisscross this area. Some are worth exploring but most are steep and hazardous. Please exercise caution on any of the paths adjacent to this segment.

**Directions to trailhead:** From Pacific Coast Highway, take Encinal Canyon Road north for nine miles (make sure to follow the road when it turns right instead of continuing straight on Lechusa Road). After mile marker 2.45, turn left onto dirt Clark Ranch Road (directly across from the county fire camp). Just a few seconds up the road, park in the large, dirt parking lot on the right.

If you're coming from the 101 freeway, exit at Kanan Road (aka Kanan Dume Road) and take it south for 6.2 miles, then turn right onto Mulholland Highway. In another mile, bear left onto Encinal Canyon Road, follow it for 2½ miles, then turn right onto dirt Clark Ranch Road (directly across from the county fire station). Just a few seconds up the road, park in the large, dirt parking lot on the right.

(Please note that if you are considering starting at the opposite end, there is virtually no parking available at the Mulholland Highway trailhead, and what little parking exists should be reserved for hikers heading westbound on Segment 17.)

From the north end of the parking lot, dirt Clark Ranch Road heads uphill to the right, while the single-track Backbone Trail heads left. A sign at the trailhead[1] reads ENCINAL TRAILHEAD - MULHOLLAND TRAILHEAD 1.1 MILES.

**The hike:** Your hike starts off with very little incline as it parallels a streambed away from the parking area. Your easy walk is surrounded by plenty of scrub and low chaparral but no shady overhang. In the spring, look for blooming yucca plants off to the sides.

Within three or four minutes, the trail crosses the streambed and starts heading west. Within a few more minutes it begins a series of gentle switchbacks as it climbs an escarpment covered with chamise.

At .3 miles, a ridiculously steep footpath crosses the trail,[2] heading up the slope to eventually join the ridgetop path described later (see alternate route on page 151). Avoid this terrible "shortcut."

It's as you approach the half mile mark that you will get your final views of the parking lot off to your left and down the mountain. However, it isn't until you are coming back down that

*Miniature stop sign and full-sized author at trail's end*

these views impress as to how high up you've climbed from Encinal Canyon Road. Within another minute, the trail turns right, leaves the escarpment, and starts paralleling Upper Trancas Canyon northward. The gentle incline flattens and from here onward, your views will mainly be of the quiet little canyon on your left.

Avoid a pair of false paths going off to the left shortly before the half mile mark.[3] About a minute beyond, you'll pass a very steep path to the right[4], sometimes signed as COMMANDO, that heads up to join the ridgetop path. As the name suggests, don't mess with it.

For the next several minutes, little seems to change. The trail maintains its nearly flat pace through sun-exposed low chaparral, maintaining a position about halfway up from the bottom of the canyon. Throughout this section you'll enjoy occasional views down into it, with lusher vegetation below. To your right, it's ridge after ridge of thirsty-looking ceanothus.

At mile .6, on your left you'll pass a flat, table-like rock that seems tailor-made for sitting. Not long after (²/₃ mile), avoid a steep side path heading up to your right that leads nowhere.[5]

A modest side trail[6] heads sharply up to the right at about .8 miles; this is the terminus of the ridgetop path (see alternate route described later). Continue straight. Within a couple of minutes, the trail crosses a small seasonal stream, with no chance of getting wet feet.

At about one mile, you will exit the top of the canyon and pass a sign facing the opposite direction that

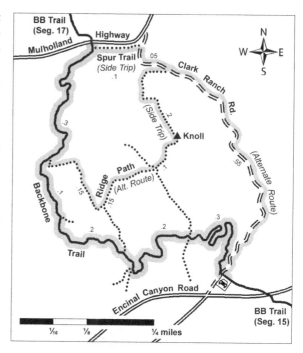

reads BACKBONE TRAIL, MULHOLLAND TRAILHEAD, ENCINAL TRAILHEAD 1.1 MILES. Mulholland Highway becomes visible and audible up ahead. Here you might be inclined to walk straight towards the highway on a gravel driveway[7] (all that remains from a house which once stood here), but instead stay on the single-track Backbone Trail, which angles to the right.

Within a minute another sign announces MULHOLLAND TRAILHEAD, ETZ MELOY MOTORWAY, 5 MILES ROUND TRIP. Oddly, this sign is not at the trailhead where the Backbone Trail crosses Mulholland Highway – it's still about a minute's walk from the trailhead. At about 1.1 miles and just before you reach the highway, the trail skirts a tiny copse of pines. This handful of conifers is all that is left of a former mini-forest here of tall evergreens which were weakened in the recent droughts and finished off by the Woolsey Fire. A side path here leads off to the right.[8]

**ALTERNATE ROUTE:** *If you want to make this hike an easy loop, shortening about .4 miles off your round trip, take the spur path to the right which parallels Mulholland Highway for ¹/₁₀ mile. At the end of the path,[9] turn right onto dirt Clark Ranch Road and follow it for another .6 miles back down to the parking lot,[1] where it rejoins the Backbone Trail.*

*The route, fully sun-exposed but not unpleasant by dirt road standards, descends moderately along the east side of the mountain, offering views over the Malibu Golf Club area and east across the broad Trancas Valley towards Buzzard's Roost as far as Castro Peak.*

*SIDE TRIP: Wish you had an ocean view on this hike? A round-trip detour of about 20 minutes, including a brief bit of huffing and puffing up 100 feet of steep elevation change, will take you to the top of a knoll with a superb 360° panorama of the Trancas area. Looking south from the top you can see waves breaking on Zuma Beach at the*

*Zuma Beach below Trancas Canyon (from knoll on side trip)*

*mouth of Trancas Canyon. To the east is an eagle-eye view over the defunct Malibu Golf Club, and to the left of it, the cliffs and pinnacles of Rocky Oaks Park. To the north you can see Etz Meloy Motorway on top of the ridge, as well as the Backbone Trail working its way up to meet it. To the west loom the impressive crags of Boney Mountain and Sandstone Peak.*

*To reach the knoll, head right on the spur trail mentioned above for ¹/₁₀ mile, then turn right onto Clark Ranch Road.[9] About .05 miles down the road, veer onto a small path[10] that leads sharply to the right, heading steeply up some railroad tie steps. The first half of this path climbs quite steeply. When you get to the top of the climb, the trail follows a flat, sun-exposed ridgetop through chamise, sage and yucca. You'll reach the top of the knoll .35 miles from where you left the Backbone Trail. The casual hiker should turn around here.[11]*

**TO CONTINUE AS AN ALTERNATE ROUTE:** *If you desire a more adventurous return trip, you can continue on this path as an alternate route back to the Backbone Trail. The mostly sun-exposed path along the ridgetop offers wonderful views but has steep sections that are rutted and rocky, making it not appropriate for some.*

*To continue, follow the path beyond the knoll downhill across a saddle, then up to a second knoll (.45 miles from where you left the Backbone Trail), which is topped with the foundation of a long-gone cabin. At the three-pronged junction[12] atop this knoll, take the middle path, which continues hugging the crest of the ridge. (The path to the right peters out and the one to the left is the "ridiculously steep footpath" mentioned earlier. Avoid.)*

*At a "Y"-shaped junction[13] (.6 miles), follow the larger path that heads to the right. (Avoid the left fork, which descends to the Backbone Trail via the aforementioned "Commando" but becomes insanely steep.)*

*From here the path descends via two short but steep and potentially slippery downward pitches. It ends at the ¾ mile mark, rejoining the Backbone Trail[6] only a few minutes south of where you left it.*

From the spur trail junction, the Backbone Trail heads straight, emerging onto Mulholland Highway[14] almost immediately. An amusingly small stop sign is placed here facing the trail. To hikers it may appear like a child's toy or some kind of a gag, but it's meant mainly for equestrians and bicyclists because Mulholland Highway curves tightly in both directions and some drivers treat it like it's the Grand Prix.

You've reached the end of this segment and have come 1.1 miles. From here you can continue on to Segment 17 of the Backbone Trail across the highway, or return the way you came.

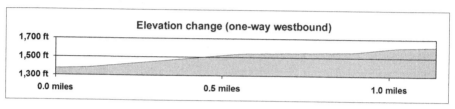

GPS COORDINATES

1. N34° 5.140', W118° 51.768'
2. N34° 5.195', W118° 51.831'
3. N34° 5.139', W118° 51.943'
4. N34° 5.159', W118° 51.951'
5. N34° 5.219', W118° 52.087'
6. N34° 5.316', W118° 52.107'
7. N34° 5.463', W118° 52.068'
8. N34° 5.488', W118° 52.027'
9. N34° 5.490', W118° 51.935'
10. N34° 5.460', W118° 51.890'
11. N34° 5.351', W118° 51.864'
12. N34° 5.300', W118° 51.904'
13. N34° 5.243', W118° 52.004'
14. N34° 5.495', W118° 52.030'

# Mulholland Highway to Yerba Buena Road
## ("Etz Meloy Motorway")

BBT segment length: 4.4 miles
Day hike length: 4.4 miles one-way
Suggested day hike: east to west, out-and-back or one-way

Elevation gain westbound: 922 feet
Elevation gain eastbound: 475 feet
Difficulty: moderate – Shade factor: 10%

Bikes allowed – Horses allowed – Dogs allowed on leash

East access: N34° 5.502', W118° 52.033'     West access: N34° 6.420', W118° 53.756'

Etz Meloy Motorway is a dirt road, named after early homesteaders, that runs atop the lofty ridge northeast of Yerba Buena Road. We don't know much about Etz Meloy, but he sure had a cool name. And as dirt roads go, he has a pretty impressive one of those too. Gravelly, steep, and inelegantly bulldozed out of the mountainside, the road nonetheless offers unforgettable views. You'll enjoy a nearly constant panorama of the Pacific Ocean, the Channel Islands, and the rolling upland of the western Santa Monica Mountains. When the road crosses over the crest, you'll also enjoy great views looking north over the Conejo Valley, including Lake Sherwood, Thousand Oaks, rocky buttes, farmland and distant high mountain ranges. One thing you won't enjoy, however, is shade.

But wait – the motorway is only the *second half* of this hike. The first half is a very different experience. The recently built single-track trail heading up the ridge to meet the motorway makes for a charming trek into the backcountry across unspoilt chaparral-covered slopes. As the trail climbs higher, expect ever-widening vistas of the Santa Monica Mountains – pastoral ranchland, pointy peaks, rugged rock spires, twisting roads, and eventually a peek at a distant stretch of Malibu beachfront.

Designed for bicycles and equestrians as well as hikers, the single-track trail's incline is so gently graded that it takes virtually the same amount of time to hike up the 650 feet of elevation gain as it does to come down it. To maintain this gentle grade, the trail switchbacks like crazy, making it twice the length of a direct route. Just the sight of these switchbacks on a map, appearing like a tangled mess of spaghetti, is enough to discourage the unenthusiastic hiker, but in reality they make the climb easy. Not surprisingly, the trail is popular with cyclists. Most of them are respectful trailmates, but be prepared to encounter the occasional klueless kamikaze.

This segment could be done as a long out–and-back trip or as a one-way trip. The latter requires that you either park cars at both termini or have someone drop you off and pick you up. You could also possibly use a rideshare service, but verify in advance that you can get a cell signal at the trailheads (see Chapter 3 for more about using rideshares).

*Panorama looking north from Etz Meloy Motorway*

**Directions to trailhead (if hiking out-and-back):** From Pacific Coast Highway, take Decker Canyon Road (Route 23) north for 4.8 miles to Mulholland Highway. Turn right onto Mulholland and drive another ½ mile. (If you're coming from the 101 freeway, exit Route 23 south, aka Westlake Boulevard. Take Route 23 south for 6.6 winding miles.) The trailhead[1] is on Route 23 just east of mile marker 5.20 and a bit west of marker 5.40.

**Parking at the trailhead:** Although the National Park Service is planning on developing a trailhead parking lot, to date there is none nearby. A few small turnouts to the west along Mulholland Highway offer a bit of parking. Your closest shot for one or two nearby spaces would be to park off of dirt Clark Ranch Road, just south of where it leaves Mulholland Highway about ¹/₁₀ mile east of the trailhead, then walk west ¹/₁₀ mile along the spur connector trail to the Backbone trailhead (see Segment 16).

If these are unavailable, you will have to park in one of the dirt turnouts near the intersection of Decker Canyon Road and Mulholland Highway ½ mile west of the trailhead, then walk back along Mulholland Highway (with little to no shoulder space).

Another option is to park in the large parking lot at the trailhead for Segment 16 and hike up dirt Clark Ranch Road, then continue on to this segment. It's more pleasant than walking along Mulholland Highway, but adds .7 miles each way plus 230 feet of elevation gain. (For directions to this parking lot, see Segment 16.)

There is a gravel parking lot near the western trailhead. If you choose to do this hike as an out-and-back from the west, simply reverse our hiking directions. We don't recommend this however, because it will involve a long trek down the mountain at the end of the trail, only to have to turn around and trudge back up – not a very satisfying hike.

**Directions to trailheads (if parking two cars for a one-way hike):** First, park your pick-up car at the end of the hike. From Pacific Coast Highway, take Decker Canyon Road (Route 23) north for 4.8 miles. Turn left on Mulholland Highway. (If you're coming from the 101 freeway, exit Route 23 south, aka Westlake Boulevard, then take it south for 7.2 winding miles, then turn right onto Mulholland Highway.)

After turning onto Mulholland, take it 0.4 miles and then turn right on Little Sycamore Canyon Road. (When you cross the Ventura County line, the road changes its name to Yerba Buena Road and becomes rougher). Two miles after you left Mulholland Highway, park in the large gravel lot on the south (left) side of the road.[2] A small sign here reads FEDERAL PROPERTY. (The western trailhead[3] is a .2-mile walk east along Yerba Buena Road, about a minute past its intersection with the paved terminus of Etz Meloy Motorway.)

Next, drive your drop-off car (or rideshare if you can get it) to the start of the hike. Head back eastbound on Yerba Buena Road. In two miles, turn left onto Mulholland Highway. In another .4 miles, you'll be at the intersection of Mulholland and Decker Canyon Roads. To reach the trailhead, turn left again onto Mulholland and drive another ½ mile. (See "Parking at the trailhead" on previous page.)

**The hike:** From the trailhead[1] at Mulholland Highway, the single-track Backbone Trail heads north, crossing a meadow bordering Lucky Ranch. Within a minute, you'll cross a seasonal stream, the first of several minor stream crossings on the lower portion of the hike.

For the first mile or so, the trail switchbacks broadly up the ridge, taking you through a pristine garden of western mountain mahogany, sage, yucca, chamise, red shank, ceanothus, scrub oak and laurel sumac. As the trail climbs, it frequently dips in and out of a series of small canyons, some of which have the feel of a tiny jungle with tall grass, reeds, and vines.

With no powerlines, roads or structures here, you may feel like you've gone out into the wilderness. Only a few artificial encroachments undermine this appearance: a bit of pastureland off to the left at ½ mile sometimes has black and white goats grazing on it, and a house or two ahead on the ridge are the only human-made structures visible anywhere around here.

1.1 miles into the hike, a switchback to the right in an area of waist-high chamise and sage offers up your first truly impressive vista of the Malibu countryside:

*Volcanic cliffs to the west, seen from Etz Meloy Motorway*

mountain peaks, rolling ranchland, and glimpses of Mulholland Highway. On a clear day your view also takes in plenty of ocean, including the double peaks of Catalina Island to the southeast, sometimes appearing to be two different islands. To the due south, you can often see pint-sized Santa Barbara Island, and if you're lucky, the low, flat profile of San Nicholas Island nearly 70 miles away to the southwest.

At 1.3 miles, you'll pass through a more arid stretch of low, scrubby chaparral and cross a brief patch of purple Sespe Formation sandstone, during which the path momentarily turns purple. From this spot you'll be treated to a wide, expansive vista looking to the south and the east. Looking from left to right are the craggy pinnacles atop Saddle Rock, then the

*Looking north on a clear day*

antenna-crowned summit of Castro Peak, then Buzzard's Roost Peak atop Zuma Ridge, as well as Lucky Ranch where you began your hike and Encinal Canyon Road winding its way up from the coast. You may also catch a glimpse of waves breaking on Zuma Beach down past the mouth of Trancas Canyon, with Catalina Island beyond.

The upcoming third of a mile may feel distinctly less than wild, due to a recently built large house which stands out prominently atop the ridge above – a rather in-your-face reminder that much of the open space adjacent to the Backbone Trail is private property and remains at risk of so-called "improvement."

By 1½ miles, you'll climb to an open space atop a small divide. Arid and sandy, this area is still scarred by the remnants of an old road that once ran from Etz Meloy Motorway down to Mullholland Highway. A few minutes later, you'll cross the disused road again. It's here that the trail finally leaves the seemingly endless switchbacks behind and angles to the west along the ridge.

A sharp right turn at 1.8 miles offers another fine view, this one including more of Zuma Beach. Then, about two miles into the hike, the trail begins working its way around the upper basin of a wide and gentle canyon. Continue around the broad canyon until the trail switchbacks right (2.4 miles) to begin its final push towards Etz Meloy Motorway.

At the switchback you'll get another good view of the area, a sort-of overview of the hike you've taken to this point: the trail meandering up around the canyon, a couple of noticeable houses, and Mulholland Highway far below. Directly to the east rise some rocky pinnacles just north of Mulholland. In the same direction, you might be able to spot 11,503-foot Mount San Gorgonio 116 miles away. To the right are Castro Peak and Buzzard's Roost Peak. The closed Malibu Golf Club lies below. Looking to the left, Etz Meloy Motorway, which you'll be joining shortly, is clearly visible higher on the ridge ahead.

It's only another five minutes of easy uphill hiking to reach the end of the single-track trail where it meets Etz Meloy Motorway,[4] a wide dirt road, at 2.6 miles. A sign here advises that Yerba Buena Road is 1.7 miles. Technically, this would be true, if you stuck to the road the whole way and then climbed over the private-property gate blocking the end of the motorway. But the recently built Backbone Trail bypass adds a good 1/10 mile to what the sign suggests.

By now you have climbed about 600 feet, but the excellent view from here of the surrounding peaks, rolling ranchland and distant Zuma Beach is only a taste of what lies ahead. Turn left onto the road, heading uphill. (A right turn east and downhill on the road quickly runs into private property.)

You'll start off by climbing up one of the steepest sections along the motorway, with slippery gravel requiring cautious footing. But in about five minutes the harsh incline relents, and a perfect panoramic view opens up to the south from a curve in the road at about 2¾ miles. From here you look out over miles of Pacific Ocean, several Channel Islands (now including Anacapa, Santa Cruz and Santa Rosa Islands to the far west), the white radar dishes of the Triunfo Pass Satellite Earth Station, and plenty of lesser ridges and canyons, all strung together by the S-curves of Yerba Buena Road and Mulholland Highway. Get used to seeing this enchanting view, as one version of it or another will accompany you for most of your trek along the motorway.

In about another five minutes, after gently cresting a local high point, you'll cross a small saddle at 2.9 miles, a spot that was private property until it was acquired by the National Park Service in 2011. The saddle offers your first view to the north looking down on the housing developments of Westlake Village and Agoura Hills, followed by ridge after distant ridge.

Where the road momentarily widens at the saddle,[5] a couple of side trails head off to the northwest and northeast. The Backbone Trail continues to follow the dirt road, which hugs the south side of the ridge.

*SIDE TRIP: For a wonderful 360° panoramic view of the entire area (and the best view on the hike), take this short side trip of 1/10 mile out-and-back, which heads up the path to the northwest (to the left).*

*The gravelly little path climbs steeply up the north side of a knoll. From the small summit, you can see in every direction for miles. Looking east and then turning right, you should be able to spot Castro Peak, then Buzzard's Roost, then the Pacific Ocean, then the white radar dishes, Mount Triunfo, Boney Mountain and Sandstone Peak. Continuing looking to the right on the north side of the ridge, you can see all the way to Mount Pinos in the distance, with Lake Sherwood, Thousand Oaks and Westlake Village down below, and finally the San Gabriel Mountains to the northeast. It's definitely worth the trip up here, and if you're seeking a "destination" to justify hiking this segment, you'll get the closest thing to it right here.*

The other side path leaving from this same point – the one heading east – is a thinly used trail that runs up and down along the crest of the ridge but slowly peters out in about ¼ mile. The views it offers are not as good as the ones from the little knoll to the west, so it isn't recommended.

From the saddle, the Backbone Trail descends steeply, offering views looking down on ranchland a good four hundred feet below, including a few houses, roads and some outbuildings. Your descent ends at 3.1 miles as you cross another

*Looking down the Arroyo Sequit as it flows to the ocean*

lower saddle,[6] again with a similar widened area of road and another brief view looking north. From here you can look down a rugged tributary canyon feeding north to Carlisle Creek. To the south, the same sweeping panoramic views of the ocean continue.

(Some online maps show a "trail" heading westward from here, hugging the ridgeline. Warning: this is not a suitable trail for anyone, as it rejoins the Backbone Trail about .4 miles to the west[7] via a ridiculously steep, dangerous and slippery descent. Do yourself a favor and avoid this hazardous route.)

Once past the lower saddle, the road heads up steeply, and after a few minutes of pretty stiff climbing, at 3.4 miles you'll make it to the high point on the hike: 2,400 feet.

Roughly a minute later (3.5 miles), where the road makes a sharp right hairpin turn across a jutting shoulder of the mountain, you will get what is probably the best southerly view on the entire hike. From this lookout you can see the same ocean panorama (yawn) that you've been enjoying during most of the hike… only far more of it. The view from here extends farther westward to include Sandstone Peak and Mount Triunfo. An old trail that once led down from here[8] to the ranch below is now overgrown and impassible. Remnants of it are still visible a few hundred feet down the mountain.

For a few minutes, the motorway heads more steeply downhill to a minor junction[7] at 3.6 miles. (The scant footpath heading north from the junction follows the bed of an old road that once descended 1,300 feet to Carlisle Road, but has now become so overgrown as to be nearly impassable, and doesn't make a suitable alternate exit.)

From here, Etz Meloy Motorway briefly straddles a flat–topped pass. To the south, the gaping canyon of the Arroyo Sequit drops nearly 2,000 feet as it heads to the sea. Directly below, Yerba Buena Road is clearly visible (and sometimes audible too), as it clings tightly to the canyon's upper slopes.

Beyond the pass, the Backbone Trail switches to the north side of the ridge, leaving the Woolsey Fire burn area for the rest of the hike and serving up a steady stream of wonderful views to the north across the Conejo Valley. For the next several minutes, you'll look out over Lake Sherwood and the farmland of Hidden Valley, as well as the community of Westlake Village, and in the far distance, the long ridge of Liebre Mountain. To the east are some blocky-looking buttes just off of Route 23.

Your most impressive northern panorama comes at about 3.8 miles, as the road rounds its first left curve after the pass and you cross the line into Ventura County. The view from here widens even more to include the arid mesas above Carlisle Canyon to the west as well as Mount Baldy and the San Gabriel Mountains to the east.

At 3.9 miles, the relatively flat walk ends as the motorway kicks off a determined descent towards Yerba Buena Road, dropping 300 feet along the northern slope of the mountain. Parts of this descent can be steep, gravelly and rutted.

After a few minutes of downhill, look for the newly built bypass trail that exits the motorway at a "Y"-junction at mile four.[9] The single-track trail branches off to the left and uphill while the motorway heads right and downhill. (If you overshoot this turnoff and continue down the motorway, you'll find yourself stuck behind an imposing gate.)

The trail climbs the ridge gently for a minute and then starts a descent

*Yerba Buena Road winding its way west from the Arroyo Sequit viewpoint*

towards Yerba Buena Road through a mix of chamise, ceanothus, red shank, manzanita and a little laurel sumac. At a small junction[10] (4.1 miles), the main trail turns right and downhill. (The slight path heading uphill to the left is what remains of the old, rough route that hikers in the know once used to bypass the gate. It climbs the ridge, then descends toward the motorway but no longer connects.)

Once past the minor junction, the Backbone Trail heads down a short but steep pitch, then switchbacks its way gently down to Yerba Buena Road, offering views looking out over the pass below.

This segment ends at 4.4 miles, where the Backbone Trail crosses Yerba Buena Road.[3] Segment 18 continues straight ahead. From here you could either turn around, head to the right to reach the trailhead parking lot[2] (.2 miles down the road) or continue west on to the next Backbone Trail segment.

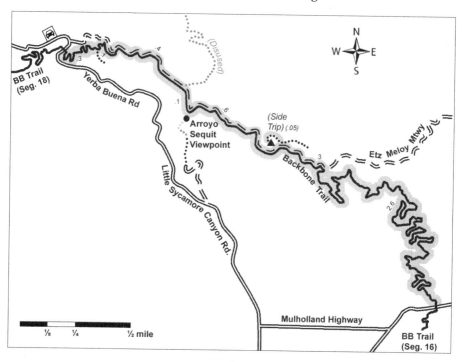

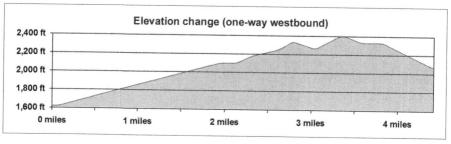

GPS COORDINATES
1. N34° 5.502', W118° 52.033'    2. N34° 6.436', W118° 53.822'    3. N34° 6.420', W118° 53.756'
4. N34° 6.008', W118° 52.519'    5. N34° 6.077', W118° 52.774'    6. N34° 6.125', W118° 52.869'
7. N34° 6.287', W118° 53.206'    8. N34° 6.200', W118° 53.186'    9. N34° 6.416', W118° 53.527'
10. N34° 6.424', W118° 53.611'

# SEGMENT 18:
# YERBA BUENA ROAD TO TRIUNFO PASS
## ("MOUNT TRIUNFO")

**BBT segment length: 4.6 miles**
**Day hike length: 4.6 miles one-way**
**Suggested day hike: west to east, out-and-back or one-way**

**Elevation gain westbound: 531 feet**
**Elevation gain eastbound: 497 feet**
**Difficulty: moderate – Shade factor: 10%**

**Bikes allowed – Horses allowed – Dogs allowed on leash**

**East access: N34° 6.424', W118° 53.762'    West access: N34° 6.848', W118° 55.105'**

"Triunfo" is Spanish for "triumph," and hats off to the folks who built this triumphant footpath, one of the newest and best segments of the Backbone Trail. Expect nearly constant ocean panoramas, impressive views of mountain cliffs, a well-graded trail and that "Wow, I'm in the middle of nowhere" feeling. Just don't expect much shade.

You'll begin with views of rugged inland mountains as you climb the side of Mount Triunfo. Then you'll cross a pass where the trail completely changes character, revealing a glorious expanse of the Pacific Ocean which will be your constant companion for the majority of the trek. Eventually, you'll cross back over the mountain, allowing some final views to the north.

The lion's share of this trail is carved out of the steep, south-facing slope of Mount Triunfo, frequently skirting cliffs and canyons and nearly always providing an awe-inspiring panorama of the Pacific Ocean, which is a mere four miles away and over 2,000 feet below. On clear days you can see Catalina Island spanning the horizon nearly 40 miles away, as well as several other Channel Islands. You'll get non-stop views looking down several canyons, which descend to a beautiful undulating plateau of chaparral, crisscrossed by Mulholland Highway as it winds its way to the sea.

As if that weren't enough, the trail makes for relatively easy hiking as well. It's extremely well graded, designed with mountain bikes in mind, and never feels very steep, climbing a relatively modest 497 feet. Portions of it are rather rocky, however, and require hiking boots.

You might want to hike on a day when the sun is not very strong or there is a decent ocean breeze because most of the trail (particularly the stretch overlooking the Pacific) is quite sun exposed. Expect only the occasional bit of shade from a rare patch of high chaparral or an overhanging canyon wall.

You'll see plenty of chamise, monkey flower and yucca in the more arid spots, and, in the slightly cooler areas, pass plenty of ceanothus, laurel sumac, and the particularly distinctive red shank with its peeling reddish bark and tufts of needles. A few minutes of elfin forest that survived the Woolsey Fire will shade you along the trail's far eastern end.

*The first part of the hike offers this sweeping view of Boney Mountain across a valley, with the ocean beyond*

This segment could be done as a long out–and–back trip or as a one-way trip if you park cars at both termini or have a friend shuttle you. (Due to the remote nature of the western trailhead, relying on a rideshare service is not recommended). If you choose to hike this trail one-way, we suggest doing it from west to east as the views tend to be better going this direction. Either way you travel it, however, the experience is wonderfully Southern Californian.

**Directions to trailhead (if hiking out-and-back):** From Pacific Coast Highway, take Decker Canyon Road (Route 23) north for 4.8 miles. Turn left on Mulholland Highway. (If you're coming from the 101 freeway, exit Route 23 south, aka Westlake Boulevard. Follow Route 23 southbound for 7.2 winding miles, then turn right on Mulholland Highway.)

After turning onto Mulholland, take it for 0.4 miles, then turn right onto Little Sycamore Canyon Road. (When you cross the Ventura County line, the road changes its name to Yerba Buena and becomes rougher). 3.9 miles after you left Mulholland Highway, park in the south (lower) parking lot at Triunfo Pass.[1] (If this lot is full, park in the north lot across the road.)

**Directions to trailheads (if parking two cars for a one-way hike):** First, park your pick-up car at the end of the hike. From Pacific Coast Highway, take Decker Canyon Road (Route 23) north for 4.8 miles. Turn left onto Mulholland Highway. (If you're coming from the 101 freeway, exit Route 23 south, aka Westlake Boulevard. Follow Route 23 southbound for 7.2 winding miles, then turn right on Mulholland Highway.)

After turning onto Mulholland, take it 0.4 miles and then turn right on Little Sycamore Canyon Road. (When you cross the Ventura County line, the road changes its name to Yerba Buena and becomes rougher). Two miles after you left Mulholland Highway, park in the large, gravel lot on the south (left) side of the road.[2] A small sign at the lot tells you you're on federal property. (Note: the eastern trailhead[3] at the end of this hike is .2 miles to the east on Yerba Buena Road.)

Next, drive your drop-off car to the start of the hike. Continue westbound on Yerba Buena Road for another 1.9 miles, then park in the south (lower) parking lot at Triunfo Pass.[1] (If this lot is full, park in the north lot across the road.)

**The hike:** From the south parking lot at Triunfo Pass, take the trail that leaves from the corner of the lot by a metal sign reading Backbone Trail, Yerba Buena Trailhead South.

The trail begins descending gently away from the road by winding down a small ravine. It won't take long until you'll feel like you're in the middle of nowhere; you won't see any electrical lines, structures, or a single dwelling of any kind. You'll be walking through a combination of chaparral and arid scrub.

For the next half mile, the trail works its way southeast along the side of a ridge, always staying just downhill from Yerba Buena Road as it winds in and out of small canyons.

As you cross a streambed in a small canyon .4 miles into the hike, you'll pass a small cliff on the left that becomes a cascade during rainy periods. Here, at 2,000 feet, you're at the low point on the segment.

The trail curves left at roughly .7 miles and begins heading gently up a more sizable canyon, in which you might encounter a brief patch of tall chaparral featuring ceanothus, toyon and red shank. The red shank makes a particularly picturesque shrub, with its twisted, peeling reddish bark and soft, needle-like leaves resembling a small pine tree. They're common in Baja but rare along much of the Backbone Trail.

*Red shank is abundant in the Triunfo Pass area*

Once the trail crosses the canyon at .9 miles, views begin opening up to your right across a sweeping valley, revealing the rugged palisades of Boney Mountain. To the left of it is a limited view of the Pacific. The taller summit to the right of Boney is Sandstone Peak, the highest point in the Santa Monica Mountains at 3,111 feet, and ironically, one of the few peaks in the entire range that is of volcanic origin and *not* made of sandstone.

After a single switchback up the canyon, you'll come to a junction at 1.1 miles.[4] The Backbone Trail curves sharply to the right by a sign that reads Federal Property.

*ALTERNATE EXIT: The path to the left is a short spur that leads, in half a minute's walk, to Yerba Buena Road east of mile marker 7.54.[5] It's your only exit to Yerba Buena Road before the Backbone Trail heads off over the mountain.*

After veering right at the junction, the Backbone Trail turns southwest, commencing a slow but steady climb of about 300 feet up the west escarpment of Mount Triunfo.

At 1.2 miles, you'll pass below a rock hoodoo that towers over the trail on the left. Its red, pink and purple sandstone is part of the Sespe Formation laid down 30 to 40 million years ago.

As you ascend farther up the ridge, the vistas of Boney Mountain across the valley grow wider and more spectacular. Even though you'll only be climbing to 2,450 feet, the hike starts to take on a high mountain feel. This section of the trail tends to be shaded until about noon, as the steep hill on the left blocks the morning sun, allowing for some areas of less thirsty-looking vegetation.

The trail makes a right turn to the northwest at 1.7 miles; from here you'll be able to see over the crest of the Santa Monica Mountains to farther, higher ranges. To the north, 8,847-foot Mount Pinos comes into view on your far left, in the same direction as the Triunfo Pass Parking Lot, which looks puny from up here. To the right, Mount Baldy and the high peaks of the San Gabriel Mountains, often snow-covered in winter, form a backdrop to the encroaching San Fernando Valley sprawl.

A single long switchback will take you to the top of the ridge where the trail heads over a small wind-swept pass at 1.9 miles. From here onward the character of your hike will change dramatically. You'll say good-bye to views of Boney Mountain and will be treated instead to nearly constant panoramas of the ocean. The trail becomes more sun-exposed, heading through south-facing low chaparral and patches of coastal sage scrub, with more manzanita and yucca than before.

About a minute's walk beyond the pass, a small, rocky path heads off from the Backbone Trail steeply uphill to the left.[6] This route is sometimes used as a "shortcut" to reach the

*Rock tower watching over the trail*

Triunfo Lookout (described later in this chapter). The very steep and crumbly path presents a major slipping hazard and isn't recommended, particularly considering that the Triunfo Lookout is accessible via the easier and far more pleasant Yellow Hill Fire Road Trail about a mile farther into this hike (see upcoming side trip description).

Continuing on the Backbone Trail, you'll soon crest the high point on the hike – about 2,450 feet – at mile two. Here the trail swings left to parallel the coastline, offering the first of many great Pacific panoramas. On a clear day you can see at least five Channel Islands, and the longer you stare, the more manage to become discernable. You can see the coastline curving south to the flat-topped hill of Palos

Verdes, once an island itself. Down below and to the right is the rugged, sheer-walled canyon of the Arroyo Sequit's west fork, better known to hikers as The Grotto. To the left, a beautiful undulating plateau separates you from the ocean, with Mulholland Highway snaking its way along its folds.

Atop the plateau, a conspicuous cluster of huge white radar dishes will prove to be a ubiquitous landmark throughout the rest of your hike. This is the Triunfo Pass Satellite Earth Station, operated by AT&T to send telephone and HDTV signals to and from satellites. Although they may look like toys from up here, each dish is over 100 feet tall and weighs 146 tons.

For the next 2¼ miles beyond the high point, the trail slowly and gently descends along the side of the ridge, winding in and out of six small but very steep and rugged canyons that cut into the mountain's southern slope and drain into the east fork of the Arroyo Sequit down below. Each canyon offers a distinctive view looking down toward the ocean and may at any time surprise you with a blast of heat or cool from its own microclimate. Between

*The radar dishes far below one of Mount Triunfo's many rugged canyons*

each canyon crossing, the trail parallels the ocean high on the ridge, offering a gradually changing perspective on the Pacific in a truly quintessential west coast view. In general, the farther you go, the less stunning the views become as you gradually lose elevation and slowly pull away from the ocean.

At 2.2 miles, you'll cross the first canyon, small but rugged. About a quarter of a mile later, as you round a major bend to the left across a jutting ridge, you'll get what are probably the most panoramic views on the entire hike, at first looking west over The Grotto and the western Channel Islands, then a few minutes later, looking southeast along the coast and over the rolling plateau a thousand feet below you.

You'll come around the head of the second canyon at 2.7 miles. Overhead are some small orange cliffs and, looking down-canyon, a rather impressive cliff that forms its southern slope.

Three miles into the hike, you'll reach a couple of large eucalyptus trees off to the left side of the trail. They look very out of place because they're so much larger than any other vegetation, and they are the only eucalyptus anywhere on the trail. A path[7] just beyond them heads to the left, up the hill behind the trees.

**SIDE TRIP:** *This path is the start of an excellent side trip to the top of the Triunfo Lookout, a 2,620-foot peak which housed a fire lookout station until the 1960s. The 360° top-of-the-world views from the summit include a panorama of the coastline, several Channel Islands, Boney Mountain, the light suburbia of Thousand Oaks, the San Gabriel Mountains, Mount Pinos and more. On a clear day you can see the San Bernardino Mountains and Mount San Jacinto near Palm Springs over 125 miles away.*

*The experience is somewhat tempered by the concrete base of the dismantled fire tower and an unfortunate amount of broken glass. This place appears to be one of those party spots where, for reasons we've never understood, people trek far out into nature to get messed up, and then deliberately mess nature up.*

*This route, a footpath that follows the old remnants of Yellow Hill Fire Road, is the longer of the two routes to the summit, but it's also the easier and more sensible approach.*

*After turning off the Backbone Trail onto the side path, follow it uphill for about a minute until you get to a "T"-shaped junction[8] with a larger trail that follows the remnants of Yellow Hill Fire Road. Turn left. (A right leads to Yerba Buena Road in about three minutes, coming out at the same location as the alternate exit described later on this page.)*

*The rocky trail climbs steadily up the north side of Mount Triunfo at a moderately steep pace through endless sun-exposed chaparral. As you continue upward, you'll get increasingly better views looking back north at beautiful little Lake Sherwood just below a rugged flat-topped mesa.*

*After about ¼ mile, the trail turns to the southwest and views of Boney Mountain ahead replace those looking back at the lake. The trail continues to climb steadily to a junction[9] at ½ mile. To reach the Triunfo Lookout, turn sharp left here and continue uphill for another .2 miles. (If you were to turn right instead, it would take you down to the Backbone Trail via the rocky and very steep shortcut described earlier.)*

*All that's left of the lookout tower at the summit*

Once past the eucalyptus trees, you will be closely paralleling Yerba Buena Road for a few minutes, and you might be able to hear the occasional car on it from your left. Soon, for a few minutes, you'll cross a patch unburned by the Woolsey Fire with lusher vegetation.

**ALTERNATE EXIT:** *A few minutes past the eucalyptus, another small path[10] to the left (at about 3.1 miles) provides an alternate exit to Yerba Buena Road, connecting in less than a minute's walk to a small gravel parking area on the south side of the road just west of the Ventura County 8.00 mile marker.[11]*

Just beyond the alternate exit, the Backbone Trail doubles back through a third, more sizable canyon with the same dramatic cliff formation on the right. From roughly this point onward, the lower land between you and the ocean resembles less of a plateau and more of a ridge forming the far side of the Arroyo Sequit.

The trail re-enters the burn zone and then curves through the fourth canyon at 3½ miles. This canyon is less dramatic than the previous two, but if you look up, you'll see a small, tilted cliff, rust-colored with white streaks, which practically overhangs the trail.

About ⅓ mile beyond, you'll pass through the top of the fifth canyon. To your left will be a sheer, rust-colored cliff, and down below to your right you'll see a small round seasonal pond at the brink of a steep drop into the canyon.

Shortly beyond, you will cross a miniature pass at mile four. (The lightly used footpath here going to your left[12] angles over the top of the ridge, then follows the overgrown remains of an old road down to meet Yerba Buena Road at mile marker 8.58, where there is limited parking at a turnout. Seeing as it doubles back considerably for .2 miles and is this close to the end of the hike, it doesn't make a good alternative exit and isn't recommended.)

Immediately after, you'll skirt the top of the last of the small canyons that cut into Mount Triunfo. Here you'll pass through a particularly scenic stretch where the trail is carved out of the steep mountainside and the land drops precipitously several hundred feet, providing some wonderful views looking down the cliff-studded canyon to the rolling plateau below and the Pacific beyond.

*Mesa above Lake Sherwood seen from the Yellow Hill Trail (side trip)*

At 4.2 miles, the trail crosses the ridge to reveal panoramas looking northward toward Thousand Oaks, Lake Sherwood and the Simi Hills. It then leaves the burn area and descends via switchbacks to a low saddle crossed by Yerba Buena Road. Designed with bicycles in mind, these switchbacks are so gentle as to be almost comical. The route seems to be wiggling back and forth just for the sake of wiggling.

The switchbacks end at about 4½ miles and the trail crosses the saddle through a shady homestretch of elfin forest, including an impressive stand of red shank and some very tall, old chamise.

In another ¹⁄₁₀ mile you'll reach the end of the trail at Yerba Buena Road.[3] From here you can turn around or continue straight across the road on to Segment 17. If you are finishing a one-way hike, your parking area[2] is a .2-mile walk up Yerba Buena Road to the left.

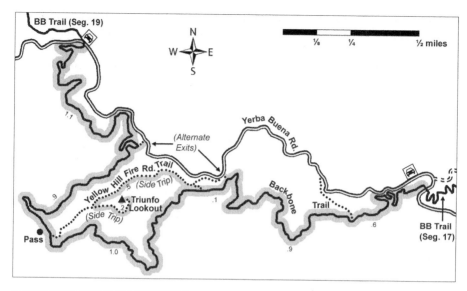

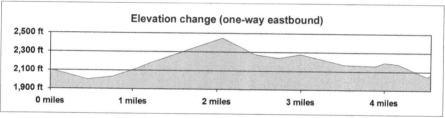

GPS COORDINATES

1. N34° 6.848', W118° 55.105'   2. N34° 6.436', W118° 53.822'   3. N34° 6.424', W118° 53.762'
4. N34° 6.549', W118° 54.863'   5. N34° 6.560', W118° 54.847'   6. N34° 6.266', W118° 55.212'
7. N34° 6.408', W118° 54.629'   8. N34° 6.422', W118° 54.663'   9. N34° 6.352', W118° 55.052'
10. N34° 6.446', W118° 54.532'   11. N34° 6.449', W118° 54.554'   12. N34° 6.317', W118° 54.051'

# TRIUNFO PASS TO BIG SYCAMORE CANYON
## (VIA BONEY MOUNTAIN)

BBT segment length: 9.9 miles
Day hike length: 13.0 miles one-way
Suggested day hike: east to west, one-way

Elevation gain westbound: 1,094 feet
Elevation gain eastbound: 3,182 feet
Difficulty: hard – Shade factor: 10%

No bikes or dogs allowed past state park boundary
Horses allowed on portions only

East access: N34° 6.875', W118° 55.093'      West access: N34° 4.305', W119° 00.860'

This hike, the granddaddy of all Backbone Trail segments, is loaded with superlatives: it's the longest, has the largest elevation change, is the most rugged, crosses the deepest wilderness, climbs the highest peak, has the widest variety of plant communities along it, and is probably the most spectacular. It's also the most challenging, but it is worth the effort. At 13 miles, your feet might hate you, but the rest of you will be grateful.

The first half of this segment takes you up over the crest of the Boney Mountain State Wilderness. If you've ever driven Yerba Buena or Las Posas Roads, looked up and seen a massive mountain with crags and jagged cliffs looking like something out of Arizona or Utah... that's Boney Mountain. Being volcanic in origin instead of the sandstone typifying most of the Santa Monica Mountains, Boney looks and feels different, more rugged. The Backbone Trail runs along its spine past the bizarre chimneys, domes and mesas which poke up like great sandcastles from its chaparral-covered highlands. You'll get the occasional million-dollar ocean view from the mountain's sheer drop-offs and you'll feel far away from civilization since you're in a bona fide wilderness preserve.

En route, you'll have the opportunity to scramble up the top of 3,111-foot Sandstone Peak, the highest point in the Santa Monica Mountains. Expect superb views in all directions except on days when fog shrouds the peak, transforming it into an eerie limbo land in the sky.

The second half of the hike heads down a lengthy 3,000-foot descent through a fascinating crash course in different environmental zones: first through a rocky alpine environment, then varying chaparral, then rolling grasslands, then a scrubby desert-like region, and finally a huge flat-bottom canyon sheltering a sycamore savannah. En route you'll have the opportunity to squeeze through Chamberlain Rock, a perfectly split boulder with a claustrophobic crevice.

After hiking for 9.9 miles, you'll have a choice to make. If you were to stay on the Backbone Trail all the way to the next car access (which is near the ocean), that would require an additional climb of 1,078 feet and another 6.4 miles. Good luck with that. Instead, we suggest that you "bail" off of the Backbone Trail at this point

*This view of Boney Mountain dropping off to the Oxnard Plain is one of the best on the entire Backbone Trail*

and walk along the nearly flat Big Sycamore Canyon Road to the nearest car access (which is also near the ocean). Instead of an additional 6.4 miles, this will add only another 3.1 miles with virtually no elevation gain: an hour and twenty minutes of easy dirt road walking.

An even easier option (but requiring tons of chutzpah) would be to beg, bribe or blackmail a friend into riding a mountain bike up Big Sycamore Canyon Road to meet you at Danielson Multi-Use Area and let you ride it out while they walk out. This turns the final 4.7 miles of the hike into a relatively easy, if somewhat bumpy, 40-minute ride down the gently descending road, essentially shortening the 13-mile hike into an 8⅓-mile one.

In any case, the length of this segment necessitates a one-way hike, requiring two cars or a friend to drop you off and pick you up. (Due to the remoteness of both trailheads, relying on a rideshare service is not recommended). We suggest hiking westbound, since the trail drops 3,000 feet heading that direction... unless you really want to test your limits by climbing the 3,000 foot, 13-mile route eastbound.

When you're done, tell your friends you hiked thirteen miles across the wilderness of Boney Mountain. They probably won't know what you're talking about, but it sounds cool, exotic and a little crazy. You're not likely to forget this hike for a long time.

Enjoying Boney Mountain does not require the full one-way hike, however. Many hikers simply climb Sandstone Peak and then turn around. Others explore the Boney Mountain Plateau and return via the Mishe Mokwa Trail alternate route (see "A Shorter Alternative"). While these variations don't complete this Backbone Trail segment, you'll enjoy some of the most impressive scenery the Santa Monica Mountains have to offer.

Whichever route you choose, carry plenty of water: There is nary a single full-size tree on top of all of Boney Mountain and the hike is mostly sun-exposed, with no water facilities or bathrooms until Danielson Multi-Use area, a full eight miles later. Also keep in mind that much of the trail is rocky, necessitating hiking boots. If you try to hike this segment in sneakers, you'll want to kick yourself with your own tattered shoes.

**Directions to trailheads (if parking two cars for a one-way hike):** First, park your pick-up car at the end of the hike in Sycamore Canyon Campground. To get there, take Pacific Coast Highway west past Malibu. After passing Mulholland Highway, continue for another five miles into Point Mugu State Park. Turn right after a sign that reads SYCAMORE CANYON CAMPGROUND 800 FEET. To park in the day-use lot[1] adjacent to the campground for a fee, follow campground signs, then turn left just after the guard booth. (The campground offers restrooms and drinking water. Fill up here as there is no drinking water at the other trailhead). You may also park along PCH for free, requiring an additional walk.

Next, drive your drop-off car to the start of the hike. Head back east on PCH for 3.2 miles, then turn left on Yerba Buena Road. Follow the road for another seven miles as it winds up the mountain. Continue past the Circle X Park Headquarters parking lot at 5.3 miles, as well as the Sandstone Peak Trailhead parking lot at 6.4 miles. At mile marker 7.00, park in in either of the two parking lots at Triunfo Pass.[2]

**Directions to trailhead (if hiking out-and-back):** From Pacific Coast Highway, turn right onto Yerba Buena Road and follow it north for seven miles as it winds up the mountain. Continue past the Circle X Park Headquarters Parking Lot at 5.3 miles, as well as the Sandstone Peak Trailhead Parking Lot at 6.4 miles. At mile marker 7.00, park in either of the two parking lots at Triunfo Pass.[2]

If you're coming from the 101 freeway, exit Route 23 south (Westlake Boulevard), take it southbound for 7.2 miles, then turn right on Mulholland Highway. Take Mulholland 0.4 miles, then turn right on Little Sycamore Canyon Road. (When you cross the Ventura County line, the road changes its name to Yerba Buena Road and becomes rougher). 3.9 miles after you left Mulholland Highway, park in either of the two parking lots at Triunfo Pass.[2]

*The Backbone Trail offers this view of Exchange Peak (left) and the 400' sheer face of Boney Peak (right)*

*Early morning fog spills over the Echo Cliffs*

**The hike:** Locate the trailhead[2] at the north edge of the northern parking lot. It's clearly labeled BACKBONE TRAIL, YERBA BUENA TRAILHEAD. MISHE MOKWA TRAIL .4 MILES. SPLIT ROCK 1.7 MILES. SANDSTONE PEAK TRAIL .6 MILES. (Note that there are actually two different "Split Rocks" mentioned on this hike. The one referenced on this sign is not on the Backbone Trail – it is on the Mishe Mokwa Trail alternate route. However, Chamberlain Rock, a major landmark ahead on the Backbone Trail, is also sometimes referred to as Split Rock.)

For the first 1⅔ miles, you'll be climbing Sandstone Peak, gaining about 1,000 feet in elevation, the vast majority of gain on this segment. You'll be on the Circle X Ranch, a large property once owned by the Boy Scouts, now administered by the National Park Service. The trail begins steeply climbing an arid ridge through a chaparral mix of chamise, yucca, red shank, ceanothus, and monkey flower. Within minutes, views open up looking south across a valley to Mount Triunfo, where Segment 18 of the Backbone Trail is visible angling up the mountainside. You might even be able to make out the white base of the old lookout tower on top.

You'll come to a junction[3] with the Mishe Mokwa Trail, which heads straight ahead, at .4 miles. (If you are on a round-trip hike and decide to return via the excellent Mishe Mokwa alternate route described later in this chapter, you'll rejoin the Backbone Trail here en route back.) At the junction, continue uphill and to the left, following a short connector trail that leads to the Sandstone Peak Trail.

After a brief climb of about three minutes, the trail descends for another two or three minutes to meet the Sandstone Peak Trail,[4] a rocky old fire road, at ½ mile.

The Backbone Trail continues straight ahead onto the Sandstone Peak Trail, resuming its uphill climb. A sign notes that Sandstone Peak is 1.1 miles ahead. As you look off to your left, you should be able to see Yerba Buena Road down below.

The trail climbs the east side of Sandstone Peak at a no-nonsense, moderate-to-steep grade through low, sun-exposed chaparral. Shortly past the junction, a sharp right switchback at .6 miles offers a wide view south over the Arroyo Sequit.

As you climb higher, beautiful rock formations begin to dot the slopes and Sandstone Peak becomes visible straight ahead, with the various mini-peaks along the ridge of Boney Mountain to the left. At .7 miles, another switchback to the right offers a better version of the earlier view over the Arroyo Sequit, now including the ocean, the length of Carlisle Canyon to your left, Castro Peak with Saddle Peak beyond it, and a host of other peaks.

About .9 miles into the hike, you'll skirt the edge of a small pass with a large rock butte towering over the trail to your left. The next few minutes beyond the butte offer wonderful views looking east down into deep, U-shaped Carlisle Canyon. Beyond the canyon, two bodies of water shimmer in the distance. The farther one is the Las Virgenes Reservoir, while the nearer one with a few islands in the middle is Lake Sherwood, adjoining the community of the same name.

After the trail turns westward around the northern slope of Sandstone Peak, an unforgettable view appears at 1.1 miles looking out across Upper Carlisle Canyon. On the opposite wall are the volcanic formations called the Echo Cliffs, popular with climbers, appearing like fondant icing rolled down the side of a cake. Above the cliffs, Balanced Rock teeters like a giant egg placed longways atop a pedestal. It's views like this that make this segment a highlight of the Backbone Trail.

In another five minutes, you'll begin a brief stint of switchbacks, climbing through a small pass with Sandstone Peak ahead and a smaller knoll behind, offering lovely views both to the northeast and to the southwest.

At 1.5 miles, the moderately steep climb finally ends as you crest the high point on the entire Backbone Trail: 3,012 feet. The summit of Sandstone Peak is a mere 100 feet higher, topping the knoll to your left. Here at the crest, a pair of crumbly-looking use paths less than a minute apart[5&6] switchback steeply up the rocks towards the peak, but why bother with either of them when the official (and much better) route to the top lies just ahead?

The trail starts to descend gently. At 1.6 miles, where the Backbone Trail turns sharp right, you'll get spectacular views looking down over the Mishe Mokwa area, including several volcanic cliffs. Here at the curve, a well-defined and signed trail heads off to the left[7] to climb Sandstone Peak.

**SIDE TRIP:** *The journey to the top of Sandstone Peak is absolutely worth the extra effort. To reach the summit, take the side path up some steps. A sign at the start reads* SANDSTONE PEAK, HIGHEST POINT IN SANTA MONICA MOUNTAINS. ELEVATION 3,111 FEET. *The steep but relatively well-maintained 1/10-mile trail takes about five minutes to reach the summit, and the farther you go, the more of a rock scramble it becomes. The trail diverges into different paths along the way, but the route tends generally up and to the left. You may have to use your hands as you near the top.*

*From the sharp-topped summit you can see... well, pretty much everything. To the south you look out over miles of Pacific Ocean, separated from you by the Arroyo Sequit and Triunfo Lookout, which you look DOWN on from here. This is one of the few places from which you can see all of the Channel Islands at once.*

*To the east, all of the Santa Monica Mountain Range lies below you, including the high points of Castro, Saddle and Temescal Peaks. To the north sprawls the suburb of*

*Thousand Oaks, crisscrossed by the 101 and 23 freeways. Beyond that, you can see all the way to the Topa Topa Mountains and distant Pine Mountain in Ventura County. To the right, you'll get an interesting perspective looking down on Balanced Rock.*

*Turning westward, your view looks out over the chaparral-covered plateau of Boney Mountain, with its rocky buttes poking up above the greenery; among them are (from left to right) Boney Peak, Exchange Peak, Tri Peaks (the highest and farthest inland with several pinnacles atop), followed by Pop Top and Big Dome. Beyond all this, the Santa Ynez Mountains near Santa Barbara form the horizon.*

*Sandstone Peak was renamed Mount Allen (to honor W. Herbert Allen, former president of the Los Angeles Council of the Boy Scouts), but most people still call it Sandstone. That's a shame, because although the vast majority of the Santa Monica Mountains are made of sandstone, this peak is not. It's andesite, a volcanic rock. Go figure. Sign the guest register just below the plaque honoring Mr. Allen and take care at the summit as the sheer drop-off from here can be a little spooky.*

After the curve, continue down the trail at a moderate descent through low chaparral. On a clear day you'll enjoy sweeping views of the Conejo Valley and Oxnard Plain sprawling in the distance. For the next two miles, you'll be traversing the Boney Mountain plateau, a strange and fascinating land of rock buttes and hoodoos, cliffs and domes, buffered by seemingly endless chamise.

You'll pass an impressive rock dome on the right at mile two. A minute or two later you'll cross a small saddle with a chimney rock on your left and a view of the ocean nearly 3,000 feet below.

From here the trail becomes rocky again as you descend gently. Good views abound south from the cliffs to the sea in the distance, and north over the rolling Boney Mountain upcountry, where the rocky outcrops seem to sprout from the scraggly groundcover.

From the saddle, the wide trail undulates gently up and down to an unsigned "Y"-shaped junction[8] at 2.3 miles. The left "path" is the remnants of the old Boney Ridge Trail, which once connected to the Circle X Ranch Headquarters about 1,000 feet down the mountain. The very steep footpath fell into disuse in the early 1990s and has completely overgrown. Surprisingly little of it remains today.

Immediately before the junction, a faint path leaves from the south and heads up through brush to the summit of adjacent Boney Peak.

*The Mishe Mokwa Trail (alternate route) high above Carlisle Canyon*

(It's a five-minute walk/bushwhack to reach the peak, but just below the top the view becomes obscured by a 20-foot pinnacle [scalable with some class III rock climbing via natural foot and handholds]. This side trip isn't recommended as you can get just as good a view from Inspiration Point or Exchange Peak, both of which are more easily accessed than Boney Peak.)

The Backbone Trail turns right, crosses a small patch of manzanita, and winds around the north slope of a rocky knoll. Then at 2½ miles, a path[9] heads off to the left by a sign reading INSPIRATION POINT.

SIDE TRIP: *It's about a one-minute scramble up to aptly named, 2,800-foot Inspiration Point: a spectacular, sheer drop-off with a terrific panorama of the Pacific coast, sandwiched in between sharp cliffs on the left and rock formations on the right. Looking north, you can see more of Boney Mountain's volcanic "bones" poking up from the green plateau, and beyond, Camarillo and the Conejo Valley. You can also see Los Angeles, the Oxnard Plain, and of course several Channel Islands. A monument here dedicated to Eagle Scout William R. Plants includes a compass that points out landmarks.*

Within another two to three minutes, the Backbone Trail comes to an unsigned "T"-shaped junction[10] at 2.6 miles. Follow it to the right.

SIDE TRIP: *Making a left at this junction will take you, in about ten minutes, up Exchange Peak. The 2,953' summit offers superb views all around, and unlike Sandstone Peak's narrow and sharp top, Exchange Peak has a relatively large and flat summit, offering space to spread out. The sheer southern precipice, one of the largest cliffs on Boney Mountain's escarpment, drops about 300 feet to the mountainside below.*

*To reach the summit, follow the side trail up a moderate incline, and after about a minute, pass two large water tanks on your right. (Exactly why these tanks are up here in the middle of nowhere is a head-scratcher.) Within another minute, the path scrambles up a small escarpment, where a crumbly-looking, narrow promontory juts off to your left far above the surrounding landscape. Past the promontory, the diminishing path splits a couple of times.[11] Follow the route that tends toward the left and uphill.*

*The puny little path continues southwest up to the base of the peak's northern escarpment. Instead of attempting to climb the cliffs there, continue westbound on the path, closely hugging the cliff base, until it emerges into a more open and rocky area and loops back eastward about .2 miles into the side trip. (Note this spot[12] on the way up. Otherwise, it can be tricky to locate on your way back down.)*

*From here, it's an easy ascent up the peak's western flank of only about a minute, requiring no rock climbing. You'll reach the summit about .2 miles after leaving the Backbone Trail.*

From the "T"-shaped junction, the Backbone Trail curves away from the cliffy escarpment and heads gently down into a chamise-carpeted upland valley.

In about five minutes (at 2.8 miles), you'll get to a junction[13] with the far end of the Mishe Mokwa Trail. Your gentle descent ends here in a little clearing (the former site of a Boy Scout campground), with the different rocky "bones" of Boney Mountain poking up in every direction.

**ALTERNATE ROUTE:** *If, at this point, you are heading home from an out-and-back hike, you could return by the Mishe Mokwa Trail. This old hiking path predating the Boy Scout ranch loops around the back of Sandstone Peak, staying far below its summit, and rejoins the Backbone Trail on the other side of the peak. It makes a superb counterpoint to the Backbone Trail as it is a totally different experience than walking the BBT along the Boney Mountain escarpment.*

*Instead of a mountaintop hike, the Mishe Mokwa Trail is a canyon trek. It descends through rugged Upper Carlisle Canyon past plenty of cliffs, crags, rock towers and the aptly named Split Rock, sampling a wide variety of vegetation along the way.*

*A warning: if you're tired and think the Mishe Mokwa Trail could make an easier shortcut back to your car than the Backbone Trail... rethink that. Mishe Mokwa is actually a tougher route because of its ups and downs and numerous boulders to scoot around along its rugged eastern half. It's very picturesque and well worth the detour, but only if you're up for it.*

*(Note that this trail doesn't allow bikes or equestrians. Also note that all mileages on this alternate route are measured from this junction. Once you finish Mishe Mokwa, you will have another .4 miles of Backbone Trail to hike until reaching the parking lot.)*

*The Mishe Mokwa Trail, which at first resembles a dirt road, climbs steadily but gently and crosses a high point ¼ mile from where you left the Backbone Trail. In another minute or two, bear right at a minor junction.[14] The drab scenery slowly improves as the increasingly rocky trail descends at a gentle to moderate grade through red shank, ceanothus and manzanita. You'll cross the main stream at .8 miles and enter the narrows of the upper canyon, an intriguing place with rock towers, ledges, cliffs and crags in every direction. The next ten minutes will offer views looking across the canyon at the Skull Rocks, a series of cliffs honeycombed with cavities and pock-marks which give the rocks a creepy, skull-like look. Some look a bit like the Sphinx; others resemble a demented Mount Rushmore.*

*The trail descends, more steeply after a while, until 1.3 miles, where a small side path[15] heads off to the left signed BALANCED ROCK. NOT AN NPS MAINTAINED TRAIL.*

*(The thinly used path to Balanced Rock meanders its way up to the rock's windswept pedestal, offering a great vista of Carlisle Canyon flanked by Sandstone Peak to the south and by Conejo Peak to the north. While a visit*

*Balanced Rock (on the Mishe Mokwa alternate route)*

to the rock may seem enticing, the path is narrow, rocky, steep and precarious in spots, with some confusing unmarked junctions along the way. This trip is definitely not suitable for children, people afraid of heights, or those in questionable physical shape. Exercise caution and judgment. If you choose to take this path, turn right (steeply downhill) at the "T"-shaped junction[16]

Split Rock (on the Mishe Mokwa alternate route)

1/10 mile into the route, then turn left at another T-junction about a minute later at the bottom of the steep pitch.[17] About five minutes later (at 1/4 mile), angle uphill to the left at a "Y"-shaped junction.[18] [The path to the right leads along the rim of the Echo Cliffs, a hazardous route with extreme exposure to the crumbly precipice, best reserved for climbers only.] Rock cairns along the path help mark the way toward Balanced Rock. After crossing a little rocky flume in a side canyon, it's a steep scramble, at times on hands and knees, up to the massive teetering boulder at .4 miles.)

Back on the Mishe Mokwa Trail, about a minute past the Balanced Rock junction, you'll come to Split Rock at 1.3 miles. The rock, which is quite obvious, sits in a wooded glade of fire-recovering oaks, bays and sycamores next to seasonal Carlisle Creek. As the name suggests, there's a huge crack in it that's easy to walk through. During the years that this was a Boy Scout camp, it was a tradition for the scouts to pass through the split, so please honor them by doing so.

At Split Rock, the trail turns sharply left, crosses the stream, and quickly becomes more rugged and dramatic. For the next mile or so, the narrow footpath rises and falls, sometimes steeply, carved into the canyon's sharp southern slope, with the Echo Cliffs dropping down the opposite wall of the canyon and Balanced Rock teetering above. The occasional fallen boulder along the trail requires some extra effort to step over or around.

Shortly into the rugged section, the trail curves around a rocky promontory at 1.6 miles; a ledge here offers a fine lookout where you can peer a couple hundred feet down into the canyon, just up from the Echo Cliffs and Balanced Rock. Climbers are often visible on the cliffs. Those with a fear of heights will find this section a little creepy.

After several minutes of this semi-rough going, the trail eventually mellows as it gradually pulls away from Upper Carlisle Canyon, descending and then re-gaining about 100 feet. Once you cross a sizable side canyon at 2.4 miles, it's only a short and gentle additional climb to a small crest where you'll rejoin the Backbone Trail[3] at the end of the 2.6-mile Mishe Mokwa Trail. To return to the parking lot, turn left on the BBT and follow it .4 miles down to the trailhead[2] at Yerba Buena Road.

From the junction with the western start of the Mishe Mokwa Trail,[13] bear left to continue westbound on the Backbone Trail. Almost immediately you'll cross a streambed and reach another signed junction[19] at 2.8 miles. The trail to the right is the first of two routes to the top of Tri Peaks, a cluster of three peaklets that form the second-highest point in the Santa Monica Mountains at 3,010 feet. The Backbone Trail heads left.

SIDE TRIP: *The memorable vista from the arid, bouldery summit of Tri Peaks is worth the climb. From the top, you look out over all of the Boney Mountain Plateau, but since it's farther inland than the other Boney summits, its best views are to the west over the checkered farmland of the Oxnard Plain and farther up the coast to Santa Barbara.*

*You can reach Tri Peaks via this side trail or by another one coming up later. Both are about the same length and the two trails can be combined to form an alternate route.*

*To take this route up, follow the rocky path as it heads toward the ridge that forms Tri Peaks. After a sharp left turn in 1/10 mile, the trail climbs steadily but moderately up the south slope. The ceanothus and red shank along this trail don't provide much shade, but they make for a slightly more pleasant hike than the other approach farther west.*

*.3 miles into the side trip, you'll arrive at a T-junction[20] where a spur trail to the summit angles to the right. (A left turn here will take you back down to the Backbone Trail via the western route described in a later side trip.) A sign here reads* TRI PEAKS *.2 MILES to the right and* BACKBONE TRAIL *.4 MILES to the left.*

*The spur trail to the summit starts off gently, ambling across a gentle valley, but within another 1/10 mile it turns left and commences a steep scramble straight up a gravelly talus slope. In another .05 miles, the trail reaches a sunbaked saddle[21] just to the southwest of the summit.*

*From the saddle, the route continues up the north peak, becoming barely discernable as it splits into a bunch of use paths that scramble up the barren, rocky slope. To reach the summit block, head towards the biggest-looking pinnacle on the far right edge of the ridge. You'll reach the summit 1/2 mile from the start of the side trip.*

*Most hikers admire the view from the southern edge of the summit and go no farther. The rest of the summit is a jumble of giant rocks that entice the adventurous with crevices, slots and small caves. Those who explore it do so at their own risk.*

*From the back side of the pinnacle, a barely discernable use path/bushwhack winds around the north flank of the peak through the maze of rocks, past crevices and a small cave, and through a miniature rock tunnel. It emerges from the boulders and chaparral to descend 2,000 feet to Satwiwa Cultural Center in Newbury Park via an unmaintained trail, passing Pop Top, the Danielson Monument & Sycamore Canyon Falls en route. The trail is very steep in spots and makes for a challenging hike that is not for beginners.*

Once past the junction, the Backbone Trail turns southwest, heading barely uphill through an isolated highland valley. It then emerges into another clearing, offering views of the rock outcroppings atop Tri Peaks on your right and Exchange Peak on your left.

The trail continues through the chaparral, surrounded by strange pinnacles, outcrops and buttes, which feel like sentinels watching over the valley. In reality they are volcanic intrusions that were forced into the older sandstone rock about fifteen million years ago. When the softer sandstone eroded away over millennia, these much harder volcanic formations remained pointing skyward.

3.2 miles into the hike, you will come to another junction.[22] The Backbone Trail continues straight ahead onto a footpath also known as the Chamberlain Trail. (What a fabulous name for a trail... of course we are not at all biased!)

*Looking down at the Pacific from the trail, with Boney Peak on the right.*

This remote path, built about thirty years ago by the California Conservation Corps as part of the Backbone system, immediately leaves the Circle X Ranch and crosses into the Boney Mountain State Wilderness (part of Point Mugu State Park). A sign here warns that there are no bicycles or dogs allowed in the wilderness area.

The trail to the right at the junction heads to the top of Tri Peaks via the western approach.

> **SIDE TRIP:** *The western route up Tri Peaks follows a more arid course than the eastern route, before merging with it and ascending the peak via a common trail to the summit. (For a more detailed description of the top of Tri Peaks, please see the side trip box on the facing page).*
>
> *From the Backbone Trail junction, follow the Tri Peaks Trail as it heads northbound, keeping right at two different junctions with use trails (in .05 miles[23] and .15 miles[24] respectively).*
>
> *The trail then steepens as it climbs through fully sun-exposed chamise up to a saddle between Tri Peaks (to the north) and a knoll with a humongous boulder atop it (to the south). Along the way, it becomes quite rutted and rocky.*
>
> *At .4 miles, you'll reach a junction.[20] To the right, the eastern route descends back down to the Backbone Trail, while a spur trail to the summit heads straight. A sign here points out that Tri Peaks is .2 miles ahead.*
>
> *The summit spur trail crosses a valley of chaparral, then in another $\frac{1}{10}$ mile hits a steep scramble up a gravelly talus slope. In another .05 miles, the trail reaches an arid saddle[21] just to the southwest of the summit. From the saddle, the route continues up the north peak, becoming barely discernable as it splits into a bunch of use paths that scramble up the barren, rocky slope.*
>
> *To reach the summit block, head towards the biggest-looking pinnacle on the far right edge of the ridge. You'll reach the summit at .6 miles from the start of the side trip.*

From the junction, the BBT returns to the ocean-facing escarpment but never actually crests the top of the cliffs, allowing for only a few brief spectacular views of the Pacific through gaps in the ridge.

At 3.3 miles, the trail tops a narrow pass with imposing volcanic ledges high on the right and pinnacle-framed ocean views on the left. Looking back you'll get your last view along the spine of the Santa Monica Mountain Range, with peak after peak visible,

*This remote lookout (side trip) offers an eagle's eye view of Boney Mountain's highest cliff and the ocean beyond*

culminating in the "M"-shaped outline of Saddle Peak. The trail becomes extra rocky as you cross into another valley, narrower than the one you're leaving.

After heading gently uphill through the little valley, you'll top another small pass at 3½ miles in an arid, open area surrounded by chamise and a few red shanks. A huge table rock that looms overhead looks like something a giant could use as a sitting stool. Looking straight ahead to the west, for the first time you can see the sprawling Oxnard Plain checkered with farmland and the sea beyond. This spot may seem insignificant, but it marks the beginning of your 2,700-foot descent to the ocean.

Less than a minute beyond the pass, a short but very steep bushwhack[25] heads straight while the Backbone Trail turns right.

**SIDE TRIP:** *The rough, overgrown path up the chamise-covered ridge will take you, in about two challenging minutes, to the edge of a 150-foot cliff with a panorama looking southward toward the ocean. From the ledge, you'll get a dramatic view to the far right of Boney Mountain's highest and most inaccessible cliff, a nearly sheer drop of over 500 feet. A spaghetti-tangle of dirt roads winds across the rolling ranchland below.*

*Here, in this soaring and remote place of wild beauty, a plaque commemorates pilot Charles "Red" Janisse – famed for skillfully navigating an ill-fated test flight of the AVE Mizar (the "Flying Pinto") in 1973 – and his wife Eleanor.*

*Note the spot where you emerged from the brush onto the bluff[26] – it can be hard to locate when leaving. Also, exercise extreme caution as these crumbly conglomerate rock ledges are exposed to a precipitous drop. Keep a respectful distance from the edge, and if you're afraid of heights or have unsure footing, don't even think of venturing to this place.*

The Backbone Trail begins its descent down the northwest rim of Boney Mountain. A few minutes into the descent, a break in the foliage at 3.7 miles[27] offers a vista not to be missed, an unforgettable view looking out over the Oxnard Plain and the Pacific coast as far as the Santa Ynez Mountains above Santa Barbara. It's a stunning sight: a huge mass of rock swooping up from the flats all the way to the heights of Boney Mountain, with its volcanic cliffs hued in brown and red, cresting like a great stone wave crashing over Camarillo.

After about ⅓ mile of descent, you'll cross a mini-landslide, including a few boulders which make good sitting rocks. Later, as the trail turns southeast into a canyon at 4.1 miles, you'll pass a fine viewpoint offering an unobstructed panorama looking north across the Conejo Valley, to the west across the Oxnard Plain and Santa Ynez Mountains, and to the south over the ocean and the Channel Islands. Things drop off sharply here a few thousand feet down to the plain. This place feels like the end of the Santa Monica Mountains.

Past the view, the trail curves into a long, gently sloping canyon with an impressive palisade at the head. In a few more minutes, it turns right as it crosses the canyon, then starts descending steeply along the canyon's southern slope. The descent is rocky and gravelly, risking a slip or two, and the trail is mostly sun-exposed, making this one of the toughest stretches of the hike.

At 4.3 miles, a confusing sign tells you that you're entering Point Mugu State Park. But didn't you already enter it back at the Boney Mountain Wilderness? What's going on is that you've now *re-entered* the state park after passing through a short stretch owned by the National Park Service.

Once past the sign, the trail drops at a moderate-to-steep pace. As you leave the canyon, the steep drop-off on the right offers more excellent views looking down on the Camarillo area. Views also open up to the west of the ocean and Channel Islands – different ones than you normally see from the Backbone Trail. These western islands – Anacapa, Santa Cruz, Santa Rosa and San Miguel – are actually an extension of the Santa Monica Mountain Range, geologically speaking.

The trail begins switchbacking down the mountainside, offering panoramas looking south over the ocean and north to the Santa Clara River Valley. At 4.6 miles, you'll get the best of these panoramic views, a wide sweep of mountains and valleys which features the vaguely heart-shaped Chamberlain Rock right in the middle.

*This "giant's sitting stool" rises about 200 feet*

About 4.7 miles into the hike, you'll reach Chamberlain Rock[28] (sometimes known as Split Rock), one of the most obvious landmarks along the hike. It's a huge conglomerate boulder bisected almost perfectly by a split, appearing to be literally cracked in two.

*SIDE TRIP: Chamberlain Rock is a fun spot, and definitely worth a break. The slot bisecting it is just thick enough that one can squeeze through, but thin enough that it's a little claustrophobic. A plaque on the rock memorializes Henry Chamberlain, the former owner of this property and a wealthy rancher who'd struck it rich prospecting for gold in Mexico. He helped in the preservation of the Circle X Ranch, and as the plaque tells us, he "loved these mountains."*

*The rock makes an excellent choice for a picnic or rest spot, particularly because the trail descending beyond this point (an elevation of about 1,970 feet) rapidly becomes hotter and more arid.*

The Backbone Trail continues switchbacking below Chamberlain Rock. Things look increasingly drier, with more grassland and scrub amidst the occasional rock outcropping. The views looking down on the fields and over the ocean remain superb. Looking back up the hill, the split in Chamberlain Rock appears particularly obvious.

At about 5.1 miles, the trail switchbacks in and out of a small canyon. These wide switchbacks last about ⅓ mile, after which the trail continues dropping moderately down the side of a ridge, in and out of sun.

The trail tops the ridgeline at 5.9 miles, then descends further by following its crest. You'll get pastoral vistas looking south over the grassy plains of Serrano Valley 700 feet below, and north across the Potrero Valley to distant 7,514-foot Reyes Peak in north Ventura County.

After roughly ten minutes of descent just below the ridgeline, the trail makes a single sharp right switchback

*Chamberlain Rock, with its obvious split, makes a welcome landmark*

and descends another minute to a T-junction[29] at 6.3 miles.

At the junction, the Chamberlain Trail meets the Old Boney Trail (a wonderfully corny name), but the route to the right is also the Backbone Trail. (To the left would take you down a more southerly route into Big Sycamore Canyon, shaving about a mile off your hike to the ocean, but we don't recommend it because it bypasses Danielson Multi-Use area with its water, restrooms, and picnic facilities.)

Bear right at the junction and descend steadily down a tributary canyon which eventually will take you into Blue Canyon. If by the name you were expecting aquamarine waters or swaying blue spruces, you'll be disappointed. This side canyon is the most arid part of the entire segment, and the more the trail descends, the drier and hotter it becomes, with vegetation growing desert-like and sparse. You've entered a different environmental zone with more cactus and scrub than chaparral. The fairly steep trail becomes gravelly and sandy, slippery at times; portions of it can get rutted and overgrown due to its relatively light use.

By 6½ miles, a small switchback takes you down a talus slope to cross the stream, which is dry in all but the

*Chamberlin (the author) meets Chamberlain (the rock)*

rainiest conditions. You'll continue to descend under full sun exposure, closely paralleling the streambed, until you cross back over it about ⅓ mile later.

It isn't until mile seven (roughly five minutes past the last stream crossing) that you'll start seeing full-size oak trees along the trail, a sign that you're about to enter Blue Canyon. You're also transitioning out of the burn footprint of the Woolsey Fire and directly into another one from an older fire, the 2013 Springs Fire, which you'll remain within for rest of the hike. If you turn around in this general vicinity and look back up the mountain, you may be able to spot Chamberlain Rock perched high on the ridge, looking hardly any larger than a pebble from here. Although it appears to be at the top, it's really only about halfway up Boney Mountain.

The trail crosses the streambed one last time over an old rusty pipe, then heads briefly over a rise to cross Blue Canyon's stream, with its bluish-grey boulders, at about 7.1 miles.

Blue Canyon is a mostly sun-exposed place, but there's a little more shade here than you had before, and in wet periods you can hear water gurgling down in the stream. Descend gently along the canyon through chaparral until you reach another junction[30] at 7.3 miles.

Here the Old Boney Trail turns right and the Backbone Trail angles left, now following a route also known as the Blue Canyon Trail. (If you were to head to the right, the Old Boney Trail would take you back up out of the canyon and eventually to Wendy Drive in Thousand Oaks.) An arrow points your way to the left with a sign: DANIELSON MULTI-USE AREA .8 MILES.

From the junction, the trail continues barely downhill through chaparral with a bit of occasional shade. Just about the time you're thinking this place should be renamed "Brown Canyon," oaks and sycamores start to appear more frequently. As you walk beside the creek bed at about 7½ miles, note the beautiful bluish grey rock and you'll see where the canyon actually gets its name. Within another five minutes, you'll start passing the remnants of an old pond on your right, now just an overgrown flat.

Once past the dam remnants at 7.9 miles, you'll cross the stream three more times, then pass through what's left of an old ranch gate as you approach the relative "civilization" of Danielson Multi-Use Area. At 8.1 miles, you'll leave the state wilderness boundary and enter the multi-use area, basically a group campground with well-needed restrooms, potable water, picnic tables, equestrian facilities, and shady sycamores.

Walk across the campground until you pass the state park maintenance manager's home on the left. This Mediterranean-style house was once the home of Richard and Molly Danielson, the owners of a large part of Big Sycamore Canyon, who sold the property to the state in the 1970s. Just past the house at 8.3 miles, turn left on Big Sycamore Canyon Road[31], which is paved for a brief stretch.

Here the Backbone Trail again dramatically changes character. It no longer feels anything like a mountain trail. Instead, it will take you on a dirt road along the long, wide, flat bottom of Big Sycamore Canyon.

Big Sycamore Canyon is an unusual place. Instead of the typical V-shaped canyon which you find everywhere in these mountains, this one is U-shaped, meaning that the bottom of it is remarkably wide and flat. In many mountain ranges, this U-shape denotes a glacier-carved valley, but here it is actually due to flood sediment.

The canyon bottom supports a rare plant community: a sycamore savannah. The plain is covered with dry grass and thistle, dotted by the occasional sprawling sycamore or oak tree. If it weren't for the yucca and chaparral-covered canyon slopes, you might at times think you were

*Big Sycamore Canyon Road*

strolling down a country road somewhere in the Midwest. Depending on your mood, you might find this place a fascinating change of pace or you might find it hopelessly tedious. It's flat as a pancake down here; certainly not much of a "backbone."

Although this place is a sycamore savannah, that doesn't mean you'll be enjoying frequent shade. The sycamores and oaks are only sporadic and few of them actually overhang the route. Walking this road is like taking the sunburn express. Also, don't expect to be alone. Big Sycamore Canyon is quite popular with hikers, bicyclists, equestrians and folks just wandering up from the campground.

After turning left onto the paved road, you'll quickly come to a "Y"-shaped junction[32] where the Backbone Trail angles off to the right onto unpaved Big Sycamore Canyon Road.

Just a few minutes later (8½ miles), continue straight at another junction[33] where a dirt road (Ranch Center Fire Road) heads off to the right.

> **ALTERNATE ROUTE:** *If you find dirt roads tedious, consider taking the Two Foxes Trail, which closely parallels the Backbone Trail/Big Sycamore Canyon Road by no more than a tenth of a mile on the other side of the stream. Unlike the flat fire road, this single-track trail negotiates gentle ups and downs through a mix of scrub and grassland on the opposite side of the canyon. It's not technically the Backbone Trail, but it's only ¹/₁₀ mile longer at 1.4 miles, with a small amount of elevation change, and it's a REAL trail. (It is very popular with mountain bikers however, especially on weekends, so if you don't like vehicles whizzing past you on a single-track path, you might want to stick with the Backbone Trail.)*
>
> *To take this alternate route, turn right onto Ranch Center Fire Road, follow it across the stream, and in about a minute (¹/₁₀ mile from where you left the Backbone Trail,) turn left at the four-way junction[34] onto the Two Foxes Trail.*
>
> *After .3 miles, continue straight where a footpath angles left[35] (the footpath quickly reconnects with the Backbone Trail). Later, at .9 miles, head straight again at a junction[36] where the Coyote Trail angles to the right and up the canyon's west side. Another minor footpath leaving to the left at 1.1 miles goes no farther than the streambed.*
>
> *The Two Foxes Trail ends in 1.3 miles at a "T"-shaped junction with Wood Canyon Fire Road.[37] To rejoin the Backbone Trail, simply turn left here and walk another ¹/₁₀ mile until the road ends at Big Sycamore Canyon Road.[38] Turn right, and you'll be back on the Backbone Trail.*

Continuing from the Backbone Trail junction with Ranch Center Fire Road, after several minutes of dirt road walking, you'll pass a small footpath[39] at 8.8 miles that heads sharply back to the right and crosses the stream, connecting with the Two Foxes Trail. Big Sycamore Canyon Road (aka the Backbone Trail) continues straight.

About ten minutes' walk later (9.2 miles), at the top of a slight rise in the road, the Old Boney Trail comes in from the left,[40] having veered off from the Backbone Trail up near Chamberlain Rock. In case you didn't fill up on water back at Danielson Multi-Use Area, there's a fire hydrant here with a water spigot on it. (Many people are surprised to find this canyon is plumbed with a series of fire hydrants. You'll pass plenty of them, spaced generally at half-mile intervals all the way to the ocean.)

At 9.8 miles, Wood Canyon Fire Road (not to be confused with the Wood Canyon Vista Trail which lies shortly ahead) turns off to the right.[38] If you took the Two Foxes alternate route, you'll rejoin the Backbone Trail here. Continue straight.

*Boney Mountain seen from the south. The Backbone Trail runs along its top.*

A few minutes later, at 9.9 miles, you will come to a decision point. This is the spot where the Backbone Trail leaves the dirt road, branching off to the right on a single-track route that promptly crosses a stream.[41] A sign designates it as WOOD CANYON TRAIL. Another reads TO OVERLOOK TRAIL.

You could, of course, continue on the Backbone Trail from here, but that would require an additional climb of 1,078 feet over another 6.4 miles. So we'll call that a different segment and save it for another hike. Instead, finish this already difficult hike in the easiest way possible, by continuing down the nearly flat Big Sycamore Canyon Road to the ocean and the Sycamore Canyon Campground Day-Use Parking Lot.

The road fords the stream ten times between here and the campground. Usually these crossings are dry, but during wetter periods they can get slightly annoying as you may have to rock-hop your way across. Other than the stream crossings, the route is highly repetitive: a mostly sun-exposed, flat, dusty walk through grasslands with the occasional shade of an oak or sycamore. It's virtually impossible to get lost down here.

The first half of this exit route (until the junction with the Serrano Canyon Trail) is the more scenic half, as the canyon quickly deepens and the walls narrow, soaring up a thousand feet on both sides.

At 11.7 miles, you'll pass the Serrano Canyon Trail heading off to the left.[42] A sign says OLD BONEY TRAIL 3.1 MILES, SERRANO VALLEY 2.7 MILES. Continue straight on the fire road. From here onward, the canyon will continue to widen and become shallower, supporting more sunny and grassy plains.

The Fireline Trail goes off to the right at mile twelve,[43] with a sign that reads To Overlook Fire Road. Continue straight. In about another ten minutes, at 12.4 miles, the Overlook Fire Road itself comes in from the right.[44] Again, continue straight.

At 12.8 miles, you'll pass the Scenic Trail, also on the right.[45] It's signed Overlook Fire Road .7 Miles. A minute later, you'll finally reach the paved campground road at a walk-around gate.[46] Continue across the campground (with its water and restroom facilities) to reach either the day-use parking lot or PCH. It doesn't matter which side of the campground loop you choose; either one will take you to the same exit.

As you leave the campground and pass the guardhouse, you'll find the day-use parking lot[1] off to your right – the end of your thirteen-mile hike. If you parked on Pacific Coast Highway, another three minutes' walk will bring you there.

## A Shorter Alternative

Don't feel like hiking thirteen miles? No problem. Most of this segment's best features lie on the flat-topped plateau of Boney Mountain, relatively close to the eastern trailhead.

Many hikers simply climb Sandstone Peak and then turn around, an out-and-back hike totaling 3.4 miles. Others go as far as the western edge of the Boney Mountain plateau, exploring the various rocky peaks along the way, then return via the Mishe Mokwa Trail alternate route – a superb round-trip hike which can range from six to nine miles depending on how many side trips you include.

On the way you can visit the highest point in the Santa Monica Mountains, explore the cliffs and rock formations of Boney Mountain, enjoy amazing views of the ocean and of the Conejo Valley interior, squeeze through Split Rock, summit Tri Peaks or Exchange Peak, and return via rugged Carlisle Canyon past the Skull Rocks, Balanced Rock and the Echo Cliffs. It's a truly unforgettable hike by any standard.

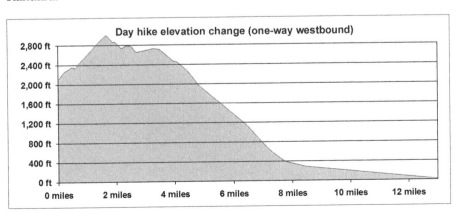

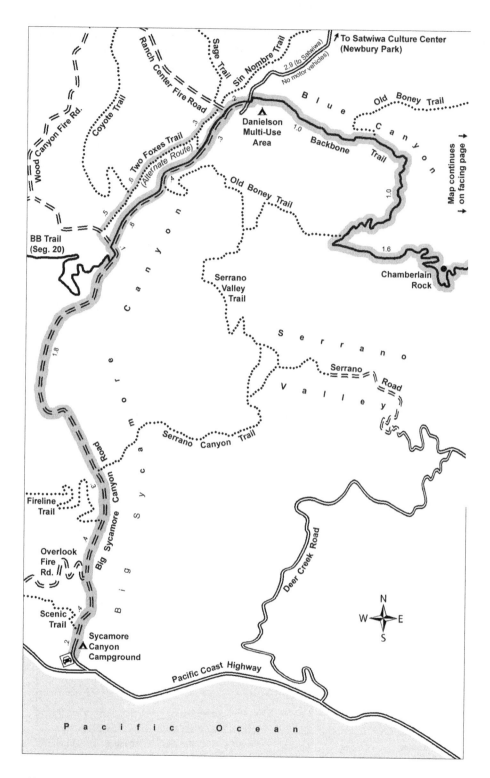

To Satwiwa Culture Center
(Newbury Park)

Map continues
on facing page

BB Trail
(Seg. 20)

Danielson
Multi-Use
Area

Chamberlain
Rock

Serrano
Valley
Trail

Serrano Valley

Serrano Road

Serrano Canyon Trail

Fireline
Trail

Overlook
Fire
Rd.

Scenic
Trail

Sycamore
Canyon
Campground

Big Sycamore Canyon Road

Deer Creek Road

Pacific Coast Highway

Pacific Ocean

N
W E
S

Wood Canyon Fire Rd.

Coyote Trail

Ranch Center Fire Road

Sage Trail

Sin Nombre Trail

2.9 (to Satwiwa)
No motor vehicles

Old Boney Trail

Backbone Trail

Two Foxes Trail
(Alternate Route)

Old Boney Trail

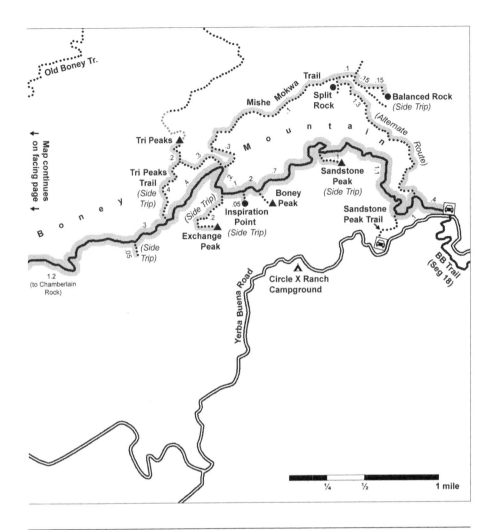

## GPS COORDINATES

| | | |
|---|---|---|
| 1. N34° 4.305', W119° 00.860' | 2. N34° 6.875', W118° 55.093' | 3. N34° 7.001', W118° 55.359' |
| 4. N34° 6.920', W118° 55.479' | 5. N34° 7.257', W118° 55.931' | 6. N34° 7.246', W118° 55.961' |
| 7. N34° 7.242', W118° 56.024' | 8. N34° 7.030', W118° 56.467' | 9. N34° 7.013', W118° 56.583' |
| 10. N34° 7.013', W118° 56.698' | 11. N34° 6.961', W118° 56.743' | 12. N34° 6.887', W118° 56.804' |
| 13. N34° 7.149', W118° 56.719' | 14. N34° 7.273', W118° 56.773' | 15. N34° 7.648', W118° 55.905' |
| 16. N34° 7.666', W118° 55.833' | 17. N34° 7.655', W118° 55.839' | 18. N34° 7.603', W118° 55.695' |
| 19. N34° 7.160', W118° 56.731' | 20. N34° 7.180', W118° 56.995' | 21. N34° 7.278', W118° 57.034' |
| 22. N34° 6.920', W118° 57.051' | 23. N34° 6.946', W118° 57.088' | 24. N34° 7.033', W118° 57.128' |
| 25. N34° 6.754', W118° 57.318' | 26. N34° 6.729', W118° 57.309' | 27. N34° 6.736', W118° 57.478' |
| 28. N34° 6.574', W118° 58.268' | 29. N34° 6.775', W118° 59.001' | 30. N34° 7.326', W118° 58.785' |
| 31. N34° 7.541', W118° 59.673' | 32. N34° 7.520', W118° 59.715' | 33. N34° 7.483', W118° 59.808' |
| 34. N34° 7.504', W118° 59.859' | 35. N34° 7.343', W119° 00.026' | 36. N34° 7.032', W119° 00.455' |
| 37. N34° 6.750', W119° 00.676' | 38. N34° 6.699', W119° 00.630' | 39. N34° 7.242', W119° 00.025' |
| 40. N34° 7.064', W119° 00.272' | 41. N34° 6.603', W119° 00.652' | 42. N34° 5.387', W119° 00.698' |
| 43. N34° 5.142', W119° 00.689' | 44. N34° 4.836', W119° 00.780' | 45. N34° 4.514', W119° 00.837' |
| 46. N34° 4.484', W119° 00.846' | | |

# BIG SYCAMORE CANYON TO LA JOLLA CANYON
## ("RAY MILLER, OVERLOOK, & WOOD CANYON VISTA TRAILS")

BBT segment length: 6.4 miles
Day hike length: 9.5 miles one-way
Suggested day hike: west to east, one-way

Elevation gain westbound: 1,256 feet
Elevation gain eastbound: 1,234 feet
Difficulty: moderate to hard – Shade factor: 5%

No bikes on "Ray Miller Trail" section (allowed on rest of segment)
Horses allowed – Dogs not allowed

East access: N34° 4.305', W119° 00.860'      West access: N34° 5.176', W119° 02.205'

For those of you who thought the entire Backbone Trail would be looking straight down on the ocean and were disappointed to find that much of it is high in the mountains or hidden in woods, this segment is for you. The first section, also known as the Ray Miller Trail, is what you thought the entire Backbone Trail would be: a spectacular single-track footpath high on the bluffs looking nearly straight down on the sea, crossing wide-open, windswept ridges of coastal scrub, cactus and yucca, with the tantalizing sound of waves crashing on the beach far below.

But that's just one third of this segment. This hike is actually a highly diverse and beautiful ramble along three very different trails stitched together to form the Backbone Trail as it winds its way through Point Mugu State Park. Each section has its own character, giving a wide variety to this hike.

After the spectacular two-mile start, you'll turn onto a fire road with easy walking along the top of a ridge. Up here, the combination of scrub and grass offers consistent sweeping views looking down 1,000 feet into tree-dotted Big Sycamore Canyon, with rugged Boney Mountain beyond, and in the other direction the wide-open grass-lands of La Jolla Valley – all with the Pacific Ocean as a distant backdrop.

Then it's time for part three. You'll head down a sandy path high on the side of a canyon hosting hillsides of spring wildflowers… and equally wild mountain bikers.

*900 feet above the Pacific Coast and Point Mugu*

*View west from Overlook Fire Road over the rolling grasslands of the La Jolla Valley Natural Preserve*

For this entire segment, there's good news and bad news. The bad news is that there's practically no shade at all. The good news is that because the vegetation grows so low, you'll enjoy nearly constant views looking either one direction or the other for the entire trek. You're rarely without something to gaze at. Just plan your trip for a cool or cloudy day.

The segment ends in the middle of nowhere: a dirt road at the bottom of Big Sycamore Canyon. But to continue eastbound along the Backbone Trail to the next vehicular access would require hiking an additional ten miles with 3,000 feet of elevation gain – pretty tough even if you're a serious backpacker. So instead, we suggest that day hikers "bail out" at this point. You could simply turn around and head home the way you came, but we – the lazy hikers – suggest heading back to the ocean on a "shortcut" along flat, dirt Big Sycamore Canyon Road. This makes the hike a one-way trip of 9½ miles, requiring you to park cars at both termini (unless you arrange a shuttle car, use a rideshare service if you can get cell access, or walk/bike back along the shoulder of Pacific Coast Highway for an additional two miles).

Another option, one requiring serious chutzpah, would be to ask a very good friend (or bribe a not-so-good one) to ride a mountain bike up Big Sycamore Canyon Road and let you ride it out while they walk out. This turns the final 3.1 miles of boring dirt road "shortcut" into a relatively easy, if somewhat bumpy, 25-minute ride down the gently descending road, requiring only a minimum of pedaling.

**Directions to trailheads:** First, park your pick-up car at the end of the hike in Sycamore Canyon Campground. To get there, take Pacific Coast Highway west past Malibu. Pass Mulholland Highway and continue another five miles into Point Mugu State Park. Turn right after a sign that reads SYCAMORE CANYON CAMPGROUND 800 FEET. To park in the day-use lot[1] adjacent to the campground for a fee, follow campground signs, then turn left just past the guard booth. The campground has restrooms and drinking water (fill up here as water is iffy at the other trailhead). You may also park along PCH for free, requiring an additional walk.

Next, drive your drop-off car (or rideshare if you can get it) to the start of the hike at La Jolla Canyon Day-Use Area, just two miles away. To do this, return to PCH and continue west for 1.7 miles, then turn right onto the small road leading up the canyon. (A sign at the entrance reads POINT MUGU STATE PARK – LA JOLLA CANYON.) Follow the road into the park, keeping left where a sign reads GROUP CAMP, and park in the La Jolla Canyon Day-Use Parking Lot[2] at the road's end (fee required). The area has restrooms and drinking water (some spigots are dry while others work – it's spotty).

**The hike:** Find the Backbone Trailhead[2] at the northeastern tip of the lot – it's the single-track path that leaves just to the right of the big gate across a fire road. A sign by the trailhead reads RAY MILLER TRAIL. OVERLOOK FIRE ROAD 2.7 MILES, SYCAMORE FIRE ROAD 5.4 MILES. (Ray, the "Guardian Of The Canyon," was the park's first camp host. The trail that bears his name and the Backbone Trail are the same path. Note, however, that the sign's mileage to Sycamore Fire Road is for an entirely different route than the BBT and doesn't apply for this hike.)

The trail heads up the ridge without delay (quickly passing a pair of side paths[3] to the group campground at 1/10 mile), and it doesn't

*Looking north at Thornhill Broome Beach and the Great Sand Dune*

take long before you can begin to see waves rolling in on the beach below. The trail maintains a gentle-to-moderate incline for the next 1½ miles or so. Considering that the Ray Miller section of the Backbone Trail climbs about 1,000 feet, it's a testament to the engineering of Ron Webster's trail crew that it almost never seems steep, allowing you to steadily chug your way up the smooth grade without breaking too much of a sweat.

For most of its climb, the trail repeats the same pleasant pattern over and over: you'll crest the top of a ridge offering a panorama of ocean, beach and mountains, then turn and head up through one of many small canyons. You'll cross the canyon through a patch of lusher vegetation, then turn back towards the ocean and up to another ridgetop viewpoint. Each repeat is higher and more dramatic than the last.

The trail climbs through coastal scrub, low and sparse vegetation with plenty of cactus and yucca, superb blooms of wildflowers in the spring… and no shade. It gives this country an open, windswept vibe, and it doesn't take long until the hike feels remote, as if you're hundreds of miles from civilization. Surrounded by open mountains, ocean and an abundance of fresh air, you can't see a single structure, powerline or any sign of man other than the trail cut into the mountain.

At .8 miles, the trail crests a ridge over-looking the ocean and makes a sharp left. From here you'll get your first of several increasingly spectacular viewpoints of the Pacific and the canyons. A small side path[4] crossing the trail here is closed to the left, while to the right ends at an overlook with a view no better than you get from the main trail.

*Boney Mountain and the fields of Serrano Valley from the top of the trail*

You'll round the top of another ridge and get an even better panorama at 1.1 miles, but it's still not as good as what's coming in a few minutes. You're looking nearly straight down on the water as several Channel Islands are visible rising from the sea. To the left you can see Thornhill Broome Beach and the Great Sand Dune, which at 400 feet is one of the tallest sand dunes on the west coast. To the right, Pacific Coast Highway winds its way under a cliff and the very top of Point Mugu peeks up from behind a ridge. Inland, the sides of La Jolla Canyon rise dramatically to a height of 1,000 feet. The trail is wide enough here to make a decent rest or picnic spot: cheap dining with a first-class view.

From this point (as well as several others on the route), the sound of waves breaking below can be impressive. The canyons act like a huge echo chamber, amplifying each crash, and the sound is extra loud because Thornhill Broome Beach is comprised of small rocks that clink together under the force of the waves, making more noise than a sand beach. The higher you climb, the longer the delay from when you see the wave break to when you hear it crash.

1.8 miles into the hike, as the trail emerges from its climb up another small canyon, it turns south to parallel the coast. Here you're about as close to the ocean as you will get on the hike, and the view looking almost straight down on it for the next several minutes is a truly unforgettable one.

After crossing yet another small canyon, you'll head out to the final point overlooking the Pacific (2.2 miles). As the trail crests the ridge, you'll enjoy your last and arguably best ocean vista, now high enough to allow a view all the way up the coast, with Point Mugu protruding into the sea. A few steps farther along the trail reveals a terrific new view looking inland towards Boney Mountain, with the grassy fields of Serrano Valley at its base.

The trail then turns sharply left and begins ascending right up the top of the ridge, allowing wonderful views of both the ocean and inland mountains.

*SIDE TRIP: Where the Backbone Trail curves left at 2.3 miles,[5] a use path heads straight and climbs the ridge, at times steeply, to a high point offering a 360° panorama of the region. At the gentle summit[6] is a mysterious pile of neatly placed stones. Is it a burial site? A monument? A boundary marker? Or just something created on a whim? Somebody knows... but not us. What we do know is that the view from here is an impressive one, including Boney Mountain to the northeast, Laguna Peak (topped with a radar station) to the northwest, and Anacapa, Santa Cruz, and Santa Rosa Islands to the southwest.*

*The tough walk to the summit takes about five to ten minutes. Beyond the summit, the path continues down, in about three or four more minutes, to the Backbone Trail, rejoining it on the left at the junction of the Ray Miller section and Overlook Fire Road.[7]*

After 2.4 miles of climbing, the trail crests a local high point. The ocean is about a thousand feet below you now, and it slowly shrinks farther behind as you make your way inland towards Overlook Fire Road, visible up ahead.

2.7 miles into the hike, the single-track Ray Miller Trail ends at a junction[7] with the Overlook Fire Road. No signs mark the Backbone Trail here, but a sign intended for westbound hikers reads RAY MILLER TRAIL TO LA JOLLA CANYON DAY-USE AREA 2.7 MILES. There is an additional metal sign on the opposite side of the road that simply says OVERLOOK, but it has been inexplicably installed backwards so that the printed part is facing away from the road. To be able to read it, you would have to be standing in the middle of the knee-high brush.

*Overlook Fire Road, part of the Backbone Trail*

Our theory: maybe the guy who hammered it in was hammered when he did it.

Turn left onto the fire road. The next two miles of Backbone Trail will be an entirely different experience than climbing the Ray Miller section. You'll be walking a country road along a ridge through a combination of scrub and grassland, with plenty of inland views and few reminders that you're near an ocean. The road makes for easy hiking, generally not very rutted, not steep on the ups-and-downs, and flat about half the time. Be prepared for more company up here as it's popular with trail runners and mountain bikers.

A few minutes past the junction, you'll get views to the right looking north towards Boney Mountain, with Serrano Valley to its right and Big Sycamore Canyon to the left.

After a brief turn toward the sea, the road resumes heading north and climbs moderately to a mini-pass at 3.4 miles. You're at 1,140 feet here, the high point on the hike. (A short use path heading to the right at the pass[8] leads to a knoll with a view no better than those from the Backbone Trail and isn't recommended.)

After the mini-pass, the road essentially flattens out, sticking firmly to the ridgetop and offering many wonderful vistas to the east. The view looking down from the sharp drop-off into Big Sycamore Canyon reveals endless fingers of small side canyons heading up the far side, plus ridge after ridge of chaparral and scrub, the occasional blooming yucca, and sporadic dots of sycamore trees down in the flat canyon plain. Looking farther north, you may glimpse the buildings of the Westlake Village/Thousand Oaks area.

At a sharp left turn at 3.8 miles, the trail switches to the west side of the ridge, revealing a whole new set of views looking out over the wide-open grasslands of the La Jolla Valley Natural Preserve framed by Mugu Peak, Laguna Peak (with its radar station atop), and La Jolla Peak. The preserve was created in 1972 to protect these pristine native grasslands from encroaching non-native invasive species. Within a few more minutes, you'll be treated to views in both directions.

By about 4.2 miles, you've descended enough to get a much closer vantage point on the La Jolla Valley Natural Preserve. More details come into view here: little paths winding through the fields, a group campground, a small seasonal pond. It's a pastoral scene, particularly beautiful in the green of springtime.

¼ mile later, your descent steepens until you reach a saddle with a "Y"-shaped junction[9] at 4.6 miles. At the junction, the Backbone Trail leaves the fire road and angles to the right onto a single-track trail also known as the Wood Canyon Vista Trail (not to be confused with the Wood Canyon Fire Road, a different route just a few minutes north of here). A small sign reads WOOD CANYON VISTA TRAIL. 2.0 MILES and BACKBONE TRAIL.

*ALTERNATE ROUTE: A popular loop hike is to leave the Backbone Trail here and return to the La Jolla Canyon Day-Use Parking Lot by descending La Jolla Canyon, with its rock-walled gorge and waterfall, making a total round-trip of 7.3 miles. While this route offers a wonderful change of scenery, it's not an easier "shortcut" back to your car. It is as sun-exposed and more rugged than the nearly flat Overlook Fire Road.*

*IMPORTANT: This route is frequently closed due to mudslides. Verify the trail status before you hike, or you'll risk being forced to turn back partway into the route.*

*To take this alternate route back, instead of following the BBT to the right on the Wood Canyon Vista Trail, continue straight on the Overlook Fire Road. In a minute or two (about ⅒ mile into the alternate route), you'll reach a 4-way junction.[10] Turn left here onto La Jolla Valley Fire Road. Follow the road downhill into the preserve, then at .3 miles angle left[11] onto a single-track trail (known as the Loop Trail) that leads through the grasslands.*

*At .8 miles, you'll pass a pond, which may or may not have water in it. Shortly thereafter (at .9 miles) continue straight on the Loop Trail past a junction[12] with another trail that heads back to the right around the far side of the pond and to a campground.*

*Continue following the Loop Trail south until you reach another junction[13] at 1½ miles. Here the Loop Trail heads back sharply to the right, leading to Mugu Peak. Continue straight at this junction onto the La Jolla Canyon Trail.*

*The La Jolla Canyon Trail gets pretty exciting as it descends into a rock-walled gorge with a little seasonal waterfall near its southern end. The canyon widens again as you approach the ocean. At 2.7 miles, you'll emerge from the canyon at the La Jolla Canyon Day-Use Parking Lot. The Backbone Trailhead,[2] where your journey started, will be on your left.*

The Backbone Trail, now following the single-track Wood Canyon Vista Trail, takes about 1¾ miles to descend into Big Sycamore Canyon along a ridge covered mostly with low, scrubby chamise. In the Springs Fire of 2013, the entire ridge burned, and what was already an arid, sun-exposed place has stayed devoid of virtually any shade. However, grasses and small shrubs have returned greenery to the area, and in the spring much of it is a virtual garden of wildflowers, sporting every color imaginable. The rest of the year, expect brown, sun-parched grass amongst the low chaparral.

The entire trail is well-graded, only moderately steep, and mostly consisting of hard-packed dirt smoothed by tire tread, making for easy hiking. Unfortunately, these conditions also make it a favorite for bicyclists who want to go *really fast* on a narrow, single-track route. Be forewarned: the Wood Canyon Vista section probably has more speeding mountain bikers than any other part of the Backbone Trail, especially on weekends. While most are responsible, you'll encounter some on this stretch who think of it as their own private thrill ride instead of the public, multiuse trail that it is.

At first, the trail rambles along the slight ridgetop, with enough sand underfoot to nearly resemble an inland dune. Within a few minutes it angles left, dropping down below the crest. An old path[14] continues straight atop the ridge but is not useable. The main trail works its way down the south side of Wood Canyon, offering nearly constant views of ever-present Boney Mountain.

At about five miles, the trail enters the first of several microenvironments that may seem surprisingly moist, at least for this area. There are plenty of reeds, some perennially green grass, and much taller chaparral here than along of the rest of the trail. The shade of the steep, north-facing slope combined with hillside seepage from springs in the area supports the unusual vegetation. You'll go in and out of these little patches for the next several minutes as you head down the canyon, which begins to offer glimpses of Wood Canyon Fire Road as well as Big Sycamore Canyon up ahead.

The trail rejoins the top of the ridge at 5⅓ miles, offering views both north and south. Within another 1/10 mile, it changes character as you make an obvious sharp right turn and begin descending the west slope of Big Sycamore Canyon. The views change: you're now looking directly out over the wide canyon, where the trail is visible up ahead winding its way down to the bottom. You can also see bits of Big Sycamore Canyon Road down below. Instead of hard-packed dirt, the trail becomes quite rocky.

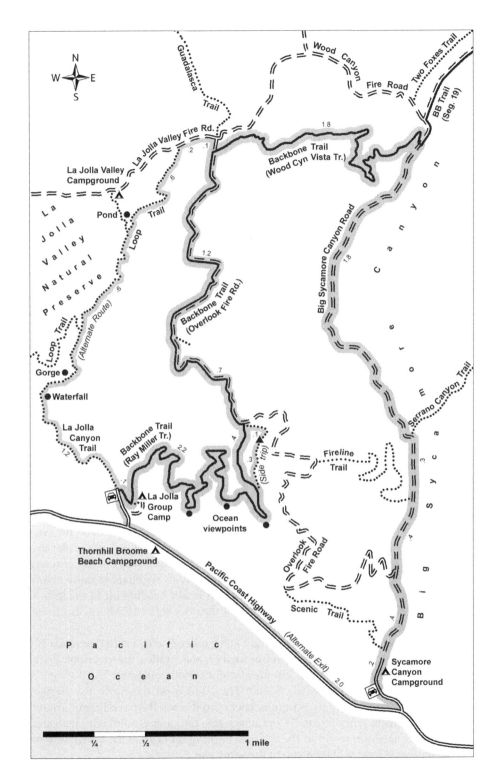

As you descend, the sycamores down below come more fully into view. At 5.8 miles, you'll make one large switchback to the north where the trail enters its final descent into the canyon down a hillside of oaks charred in the Springs Fire, some of which have resprouted new foliage.

You will make it to the bottom at 6.4 miles, where the single-track trail crosses the canyon stream and then ends at Big Sycamore Canyon Road.[15] It's a good spot to stop and refill water at a spigot a few steps south of the junction.

From here, the Backbone Trail turns left and heads north on Big Sycamore Canyon Road. But to follow the Backbone Trail eastbound to the next vehicular access

*Heading down through Wood Canyon in the springtime*

would require hiking an additional ten miles over Boney Mountain (see Segment 19), including 3,000 feet of elevation gain. Unless you're a maniac, we suggest that you "bail out" here and save Segment 19 for another day.

If you're hiking out-and-back, this is your turnaround point – time to trudge right back up Wood Canyon. But a far easier exit (assuming you've made transportation arrangements) is to turn right on Big Sycamore Canyon Road and follow the nearly flat dirt road south to the day-use parking lot at Sycamore Canyon Campground, just off of Pacific Coast Highway. Returning via this route shaves off 3.3 miles and 1,078 feet of elevation gain compared to turning around and retracing your steps on the Backbone Trail.

The walk back to PCH on Big Sycamore Canyon Road is an easy but highly repetitive one: a sun-exposed, flat, dusty march through scrubby grasslands under the rare shade of an oak or sycamore. Along the way, the road fords the stream ten times, barely noticeable in dry months and pretty easy to negotiate in rainy ones.

At 8.2 miles, you'll pass the Serrano Canyon Trail[16] heading off to the left. A sign here reads OLD BONEY TRAIL 3.1 MILES, SERRANO VALLEY 2.7 MILES. Continue straight on the fire road.

The Fireline Trail heads off to the right at 8.5 miles,[17] with a sign that reads TO OVERLOOK FIRE ROAD. About ten minutes further, at 8.9 miles, the Overlook Road itself comes in from the right.[18] Continue straight at both junctions.

At 9.3 miles, you'll pass the Scenic Trail,[19] also on the right. It's signed OVERLOOK FIRE ROAD .7 MILES. A minute later, you'll reach the paved campground loop road at a walk-around gate.[20] Continue across the campground with its water and restroom facilities to get to the day-use parking lot or to PCH. Either side of the campground loop will take you to the same exit.

As you leave the campground and pass the guardhouse, on your right will be the day-use parking lot[1], the end of your 9½-mile hike. If you parked on Pacific Coast Highway, another three minutes' walk will bring you there.

*ALTERNATE EXIT: If you've only parked one car (back at the La Jolla Canyon Day-Use Area), you can walk back to it along the reasonably wide shoulder of PCH, adding an additional two miles and making this trip a loop hike of 11½ miles total. Of course, walking along busy PCH is not our idea of fun and we don't really recommend it.*

*If you had the forethought to lock a bike here at the campground, your ride back to the day-use area along PCH will be only about ten minutes.*

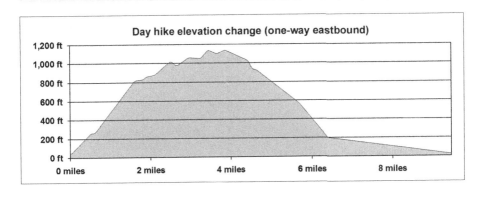

Day hike elevation change (one-way eastbound)

| | | GPS COORDINATES | | |
|---|---|---|---|---|
| 1. N34° 4.305', W119° 00.860' | | 2. N34° 5.176', W119° 02.205' | | 3. N34° 5.122', W119° 02.134' |
| 4. N34° 5.181', W119° 01.943' | | 5. N34° 5.179', W119° 01.496' | | 6. N34° 5.372', W119° 01.482' |
| 7. N34° 5.450', W119° 01.559' | | 8. N34° 5.808', W119° 01.918' | | 9. N34° 6.567', W119° 01.703' |
| 10. N34° 6.625', W119° 01.693' | | 11. N34° 6.603', W119° 01.876' | | 12. N34° 6.256', W119° 02.158' |
| 13. N34° 5.829', W119° 02.399' | | 14. N34° 6.612', W119° 01.615' | | 15. N34° 6.603', W119° 00.652' |
| 16. N34° 5.387', W119° 00.698' | | 17. N34° 5.142', W119° 00.689' | | 18. N34° 4.836', W119° 00.780' |
| 19. N34° 4.514', W119° 00.837' | | 20. N34° 4.484', W119° 00.846' | | |

# QUICK-REFERENCE TABLE

| Segment | Segment Name | One-Way Mileage (BBT Segment Only) | One-Way Mileage (Day Hike) | Westbound Gain | Eastbound Gain | Difficulty Level | Shade Factor |
|---|---|---|---|---|---|---|---|
| 1 | Rogers Trail | 7.6 | 10.2 | 2,184 | 858 | Hard | 15% |
| 2 | Musch Trail & Eagle Rock Fire Road | 3.5 | 3.5 | 248 | 1,068 | Moderate | 25% |
| 3 | Dead Horse Trail | 1.2 | 1.2 | 63 | 461 | Easy-Mod. | 45% |
| 4 | Henry Ridge Crossing | 0.8 | 0.8 | 218 | 217 | Easy-Mod. | 70% |
| 5 | Hondo Canyon Trail | 3.7 | 3.7 | 1,647 | 75 | Moderate | 70% |
| 6 | Fossil Ridge Trail | 0.7 | 0.7 | 133 | 118 | Easy | 80% |
| 7 | Saddle Peak Trail (East) | 0.7 | 0.9 | 498 | 68 | Easy-Mod. | 60% |
| 8 | Saddle Peak Trail (West) | 1.2 | 1.6 | 0 | 875 | Moderate | 35% |
| 9 | Saddle Creek Trail | 3.2 | 3.4 | 214 | 1,342 | Moderate | 35% |
| 10 | Piuma Ridge Trail | 2.0 | 2.0 | 225 | 468 | Easy-Mod. | 75% |
| 11 | Mesa Peak Motorway | 5.2 | 5.2 | 2,181 | 677 | Moderate | 10% |
| 12 | Castro Crest | 4.1 | 4.1 | 946 | 895 | Moderate | 15% |
| 13 | Newton Canyon Trail | 2.2 | 2.2 | 240 | 740 | Moderate | 20% |
| 14 | Upper Zuma Canyon Trail | 2.5 | 2.5 | 561 | 374 | Moderate | 20% |
| 15 | Trancas Canyon Trail | 2.4 | 2.4 | 360 | 708 | Moderate | 10% |
| 16 | Clark Ranch Trail | 1.1 | 1.1 | 264 | 22 | Easy | 5% |
| 17 | Etz Meloy Motorway | 4.4 | 4.4 | 922 | 475 | Moderate | 10% |
| 18 | Mount Triunfo | 4.6 | 4.6 | 531 | 497 | Moderate | 10% |
| 19 | Boney Mountain | 9.9 | 13.0 | 1,094 | 3,182 | Hard | 10% |
| 20 | Ray Miller, Overlook & Wood Canyon Vista Trails | 6.4 | 9.5 | 1,256 | 1,234 | Mod.-Hard | 5% |

| Bikes Allowed | Horses Allowed | Dogs Allowed | Toilet(s) | Piped Water | Parking Lot(s) | Roadside Parking | Picnic Table(s) | Stream Crossing(s) | Car Shuttle Needed |
|---|---|---|---|---|---|---|---|---|---|
| ✓ | ✓ | - | ✓ | ✓ | ✓ | - | ✓ | - | ✓ |
| Limited | ✓ | - | ✓ | ✓ | ✓ | - | ✓ | ✓ | - |
| - | ✓ | - | ✓ | ✓ | ✓ | ✓ | ✓ | ✓ | - |
| - | ✓ | - | - | - | - | ✓ | - | - | - |
| - | ✓ | - | - | - | - | ✓ | - | ✓ | - |
| - | ✓ | - | - | - | ✓ | ✓ | - | - | - |
| - | ✓ | - | - | - | ✓ | - | - | - | - |
| - | - | - | - | - | - | ✓ | - | - | - |
| - | ✓ | - | - | - | - | ✓ | - | ✓ | Optional |
| - | ✓ | - | ✓ | Nearby | ✓ | - | Nearby | ✓ | - |
| ✓ | ✓ | - | ✓ | Nearby | ✓ | - | ✓ | - | ✓ |
| ✓ | ✓ | ✓ | - | - | ✓ | - | - | ✓ | Optional |
| ✓ | ✓ | ✓ | ✓ | - | ✓ | - | - | ✓ | - |
| ✓ | ✓ | ✓ | ✓ | - | ✓ | - | - | ✓ | - |
| ✓ | ✓ | ✓ | - | - | ✓ | - | - | - | - |
| ✓ | ✓ | ✓ | - | - | ✓ | - | - | ✓ | - |
| ✓ | ✓ | ✓ | - | - | - | - | - | ✓ | Optional |
| ✓ | ✓ | ✓ | - | - | ✓ | - | - | ✓ | Optional |
| Limited | Limited | Limited | ✓ | ✓ | ✓ | - | ✓ | ✓ | ✓ |
| Limited | ✓ | - | ✓ | ✓ | ✓ | - | ✓ | ✓ | ✓ |

# BEST HIKES FOR...

| Segment | Segment Name | City Views | Geology/ Rock Formations | Hair-raising Hikes | History | Kids | Mountain/Canyon Views |
|---------|--------------|------------|-------------------------|--------------------|---------|------|----------------------|
| 1 | Rogers Trail | ✓ | ✓ | ✓ | ✓ | - | ✓ |
| 2 | Musch Trail & Eagle Rock Fire Road | Side trip | ✓ | Side trip | ✓ | - | ✓ |
| 3 | Dead Horse Trail | - | - | - | ✓ | ✓ | - |
| 4 | Henry Ridge Crossing | - | - | - | - | ✓ | - |
| 5 | Hondo Canyon Trail | - | ✓ | Side trip | ✓ | - | ✓ |
| 6 | Fossil Ridge Trail | Side trip | ✓ | - | - | ✓ | Side trip |
| 7 | Saddle Peak Trail (East) | ✓ | ✓ | - | - | ✓ | ✓ |
| 8 | Saddle Peak Trail (West) | ✓ | ✓ | Side trip | - | - | ✓ |
| 9 | Saddle Creek Trail | - | ✓ | - | - | - | ✓ |
| 10 | Piuma Ridge Trail | - | - | - | - | - | ✓ |
| 11 | Mesa Peak Motorway | - | ✓ | - | ✓ | ✓ | ✓ |
| 12 | Castro Crest | - | Side trip | - | - | - | ✓ |
| 13 | Newton Canyon Trail | - | - | - | - | ✓ | ✓ |
| 14 | Upper Zuma Canyon Trail | - | - | - | - | ✓ | ✓ |
| 15 | Trancas Canyon Trail | - | - | - | ✓ | - | Side trips |
| 16 | Clark Ranch Trail | - | - | - | - | ✓ | Side trip |
| 17 | Etz Meloy Motorway | - | - | - | - | - | ✓ |
| 18 | Mount Triunfo | - | ✓ | - | - | - | ✓ |
| 19 | Boney Mountain | - | ✓ | Side trips | ✓ | - | ✓ |
| 20 | Ray Miller, Overlook & Wood Canyon Vista Trails | - | Alt route | Alt route | - | - | ✓ |

| Ocean Views | Peak-bagging | Rock Scrambling | Solitude | Streams | Variety/Diversity | Waterfalls (Seasonal) | Wildflowers (Seasonal) | Wildlife | Woods |
|---|---|---|---|---|---|---|---|---|---|
| ✓ | Side trip | Alt route | ✓ | Alt route | ✓ | - | - | ✓ | - |
| ✓ | - | Side trips | - | ✓ | ✓ | - | - | ✓ | - |
| - | - | - | - | ✓ | - | - | ✓ | ✓ | ✓ |
| - | - | - | - | - | - | - | - | - | ✓ |
| - | - | Side trip | ✓ | ✓ | ✓ | - | - | ✓ | ✓ |
| ✓ | - | - | - | - | - | - | - | - | - |
| ✓ | ✓ | Side trip | - | - | ✓ | - | - | - | - |
| ✓ | ✓ | ✓ | - | - | - | - | - | - | - |
| - | - | - | ✓ | ✓ | ✓ | ✓ | ✓ | ✓ | ✓ |
| - | - | - | - | ✓ | - | - | - | - | ✓ |
| ✓ | Side trip | ✓ | - | - | - | - | - | - | - |
| Side trip | - | Side trip | ✓ | ✓ | - | - | ✓ | ✓ | ✓ |
| Side trip | - | - | - | ✓ | - | - | ✓ | - | ✓ |
| - | - | Side trip | ✓ | ✓ | ✓ | ✓ | - | ✓ | ✓ |
| Side trip | Side trip | - | ✓ | ✓ | - | - | - | ✓ | ✓ |
| Side trip | - | - | - | - | - | - | ✓ | - | - |
| ✓ | - | - | - | - | - | - | ✓ | ✓ | - |
| ✓ | ✓ | Side trip | ✓ | - | ✓ | - | - | ✓ | - |
| ✓ | ✓ | Side trips | ✓ | ✓ | ✓ | - | ✓ | ✓ | - |
| ✓ | - | - | - | ✓ | ✓ | Alt route | ✓ | - | - |

# INDEX

*Page numbers in italics refer to photographs. Number listings are at the end of index.*

Africa, 35
AVE Mizar, 180
Africa, 37
Agoura Hills, 157
Airbnb, 24
alder trees, 41, *102,* 106
Allen, W. Herbert, 174
Anacapa Island, 157, 181, 194
andesite, 37, 174
Appalachian Mountains, 14, 103
Appalachian Trail, 11, 16
Arizona, 93, 95, 98, 169
Arroyo Sequit, 34, *158,* 158, 165, 167,
    172-173
AT&T, 165
Australia, 37

Baja California, 163
Balanced Rock, 173, 174, *176,* 176-177, 187
basalt, 37, 72
bay trees, 40, 71, 78, 79, 80, 81, *82,* 82, 83,
    92, 103, 106, 108, 111, 112, 176
Bay Tree Trail, 47, 48
Bent Arrow Trail, 46
Betty Dearing Mountain Trail, 14
Big Dome, 174
Big Sycamore Canyon, 34, 36, 41, 121, 182,
    184-185, 190, 191, 194, 195, 196
Big Sycamore Canyon Road, 170, *184,* 184,
    185, 186, 191, 192, 196, 198
black bears, 42
Blue Canyon, 183
Blue Canyon Trail, 183
Blue Ridge, 14
bobcats, 32, 42
Bone Canyon Trail, 54
Boney Mountain, 18, *36,* 37, 48, 133, 134,
    144, 146, 151, 157, *160,* 163, 164, 166,
    169, *170,* 170, 173, 174, 175, 176, 178,
    *180,* 180, 181, *186,* 187, 190, *193,* 193,
    194, 196, 198
Boney Mountain State Wilderness, 169,
    179, 181, 184
Boney Peak, *171,* 174-175, *179*
Boney Ridge Trail, 174
Borna Drive, 126
Boy Scouts, 172, 174, 176, 177
Bragg, Paul, 81
Braude, Marvin, 46

Brents Mountain, 105, *116,* 121
Bulldog Motorway, 124
Butch Cassidy and the Sundance Kid,
    114, 118
butterflies, 72
Buzzard's Roost, 128, 133, 139, 144, 146,
    150, 156, 157
Buzzard's Roost Ranch, 146

Caballero Canyon, 46
Cabernet Sauvignon, 132
cacti, 4, 32, 37, 38, 40, 183, 190, 192
Cahuenga Pass, 34
Calabasas, 86, 91, 96, 102
Calabasas Peak, 94, 96, *99,* 99, 100, 103
Calamigos Guest Ranch, 23, 24
California buckwheat, 40
California Conservation Corps, 16, 104,
    179
California grizzly, 42
California kangaroo rat, 42
California mountain kingsnake, 43
California native grassland plant
    community, 37, 41, 190, 195
California State Route #23, 154, 158, 161,
    171
California Wildlife Center, 111
Camarillo, 175, 181
Cameron, James, 12
Carlisle Canyon, 159, 173, *174,* 176, 177,
    187
Carlisle Creek, 158, 177
Carlisle Road, 158
Castro Crest, 123, 126
Castro Crest Trail, 16, 123-129
Castro Motorway, 124-125
Castro Peak, 123, 125, 127, 128, 132, 133,
    150, 156, 157, 173
Catalina Island, 44, 48, 52, 85, 89, 90, 93, 94,
    99, 100, 120, 124, 125, 133, 155, 156, 160
Cathedral Rock, *46,* 47, 65, 66
cattails, 41
ceanothus, 10, 39, 48, 50, 51, 59, 80, 81, 85,
    88, 92, 96, 97, 98, 103, 104, 105, 112,
    125, 127, 137, 139, 142, 149, 155, 159,
    160, 163, 172, 176, 178
Century City, 44, 48, 49, 52, 94, 100
chalk, 119
Chamberlain, Henry, 182

Chamberlain Rock, 169, 172, 181, *182*, 182, *183*, 185
Chamberlain Trail, 179, 182
chamise, *38*, 39, 48, 49, 62, 70, 81, 96, 98, 105, 112, 116, 125, 127, 134, 139, 142, 149, 151, 155, 159, 160, 167, 172, 174, 175, 179, 180, 196
Channel Islands, 118, 153, 157, 160, 164, 165, 166, 173, 175, 181, 193
chaparral plant community, 11, 13, 17, 32, 34, 37, 38-39, 40, 48, 50, 52, 53, 54, 62, 63, 64, 70, 76, 81, 82, 83, 85, 92, 95, 96-97, 103, 106, 111, 112, 113, 114, 119, 120, 121, 123, 125, 126, 128, 133, 135, 137, 138, 139, 141, 142, 143, 144, 145, 148, 149, 153, 156, 160, 163, 164, 166, 169, 172, 174, 178, 179, 183, 184, 195, 196
chaps, 38
Chardonnay, 132
Cheney Fire Road, 65
Chicken Ridge Bridge, 44, 45, 51, *52*, 52, 55, 59
Chile, 37
chipmunks, 42, 70
Chumash people, 104, 135
Circle X Ranch, 172, 179, 182
Circle X Ranch Group Campground, 23, 24
Circle X Ranch Park Headquarters, 171, 174
Clark Ranch Road, 142, 148-151, 154
clover, 10, 41, 72, 81, 104
Coal Canyon Formation, 36
Coast Ranges, 35
coastal sage scrub plant community, 37, 40, 49, 72, 114, 149, 163, 164, 182, 183, 185, 190, 192, 195, 198
Cold Creek, 111
Cold Creek Canyon, 103
Cold Creek Canyon Preserve, 87
Commando, 149, 151
Conejo Peak, 176
Conejo Valley, 34, 65, 92, 96, 97, 98, 99, 101, 103, 104, 112, 117, 118, 153, 158, 174, 175, 181, 187
Conejo Volcanics, 37
Corpse Wall, The *98*, 98
Corral Canyon Arson Watch, 117
Corral Canyon Cave, 117
Corral Canyon Fire, 117
Corral Canyon Road, 116, 123, 124, 126
cottonwood trees, 41, 111
Coyote Trail, 185
coyotes, 32, *42*, 42, 70
crickets, 138
crows, 42

Danielson, Richard & Molly, 184
Danielson Monument, 178
Danielson Multi-Use Area, 23, 24, 170, 182, 183, 184, 185
Dark Canyon, *12*, 24, *101*, 104, 106
Dead Horse Parking Lot, 24, 69, 70, 71, 72, 74
Dead Horse Trail, 15, 16, 61, 68-73, 130
death march, 12, 26, 27, 146
Decker Canyon Road, 154-155, 161
deer, 32, 40, 42, *43*, 62, *68*, 68, 70, 123, 142
Disneyland, 27
dodder, 39, *41*, 49, 125, 128
Doors, The, 117

eagles, 42, 66
Eagle Junction, 21, 63, 64, 65, 66, 67
Eagle Rock, 48, 60, 63, 64, *65*, 65, 66-67
Eagle Rock Fire Road, 60, *61*, 63-65, 66
Eagle Spring, 66
Eagle Springs Fire Road, *15*, 47, 64, *65*, 65, 66, 75
Echo Cliffs, *172*, 173, 177, 187
elfin forest, 10, 26, 38, 78, 80, 81, 83, 88, 96, 97, 104, 106, 108, 112, 122, 160, 167
Encinal Canyon Road, 139, 140, 142, 144, 145, 146-147, 148, 149, 156
Entrada Road, 61, 69, 70, 71, 74
Equestrians, 16, 23, 54, 62, 71, 143, 152, 153, 176, 184, 185
Etz Meloy Motorway, *13*, 105, 150, 151, 153-160
eucalyptus trees, 41, 53, 54, 62, 81, 116, 165, 166
Ewen, Lois, 87
Exchange Peak, *171*, 174, 175, 178, 187

ferns, 41, 50, 58, 63, 71, 72, 78, 80, 81, 83, 92, 104, 106, 108, 112
Feuer, Margot, 14
Fire Camp #13, 142, 143, 148
Fireline Trail, 187, 198
Fish and Game Department, 42
Flintstones, The, 50
footbridges, 62, 68, 69, 70, 71, *72*, *137*, 138, 141, 143
fossils, 35, *85*, 85, 119, *121*
Fossil Ridge, 71, 76, 82, 85, 88
Fossil Ridge Trail, 16, 83, 85-89, 130
Franklin Canyon, 34
Frazier Mountain, 103, 118
Frazier Park, 88
freshwater marsh plant community, 37, 41, 81, 138, 139
frogs, 41, 43, 68, 71, 138

Garapito Canyon Trail, 46, 65
Gateway Rock, *95*, 97-98, *99*
Germany, 58
Glen Trail, 72
Goat Buttes, 96, 97, 103, 108, *112*, 112, 118
goats, 155
Google Earth, 29
Google Maps, 17, 76
gopher snake, 43
graffiti, 55, 58, 59, 85, 86, 87-88, 116, 117
grass, 26, 30, 40, 41, 50, 62, 64, 68, 72, 81, 83,
    90, 92, 104, 111, 112, 115, 119, 121, 122,
    125, 133, 135, 137, 139, 141, 143, 145,
    155, 182, 184, 185, 186, 190, 195, 196, 198
Great Sand Dune, *192*, 193
Green Peak Communications Tower, 49
Greenleaf Canyon Road, 73, 74, 75
grey fox, 42
Griffith Park, 14, 34
Grotto, The, 165

Henry Ridge, 74, 130
Henry Ridge Motorway, 76
Hidden Valley, 158
Hillside Drive, 62
Hilltop Climb Drive, 106
hollyleaf cherry, 80, 103, 112
Hollywood, 39, 52
Hollywood Hills, 14, 34, 36
Hollywood Sign, 34
Hondo Canyon, 71, 76, *78*, 80, 81, 82, 83,
    84, 86, 87, 88, 89
Hondo Canyon Trail, 16, 78-83, 87
Hondo Stream, 82, 83
hoodoos, 36, *93*, 93, 95, 99-100, 114, *164*,
    164, 174
horses, 21, 26, 52, 68, 70, 111, 116, 148, 152
Hub, The, 24, 45, 46, 47, 60, 65, 66, 67
Hubbard, Joanne, 14

igneous rock, 36, 37
Inspiration Point (in Circle X Ranch), 175
Inspiration Point (in Will Rogers Park),
    53, 54, 55, 59
Inspiration Point Loop Trail, 53-54

Janisse, Charles "Red", 180
Josepho Spur Trail, 51, 55, 59

Kanan Dume Road, 23, 124, *130*, 131, 132,
    133, 134, 135, 136, 137, 139, 142, 146, 148
Kanan Dume Road Parking Lot, 24, 131,
    135-136
Keepers Of The Earth, 75
King Gillette Ranch, 119

La Jolla Canyon, 192, 193, 195
La Jolla Canyon Day-Use Area, 24, 192,
    194, 195, 196, 199
La Jolla Canyon Trail, 196
La Jolla Peak, 195
La Jolla Valley Fire Road, 195
La Jolla Valley Natural Preserve, 41, *191*,
    195
labyrinths, 114, *117*, 117, 146
Ladyface Mountain, 97
Laguna Peak, 194, 195
Lake Sherwood, 153, 157, 158, 166, *167*,
    167, 173
landslides, 47, 78, 82, 111-12, 181
Las Flores Canyon, 88
Las Flores Canyon Road, 86, 91
Las Posas Road, 169
Las Virgenes, 104, 105, 119, 121
Las Virgenes Reservoir, 173
Las Virgenes Road, 21, 23, 102, 109, 110,
    115, 122
Las Virgenes Road Parking Lot, 24, 109,
    115
Las Virgenes Water Treatment Plant, 110,
    122
Latigo Canyon Road, 24, 123, 124, 127,
    *128*, 128, 131, 133, 134
Laurel Canyon, 34
laurel sumac, 39, *40*, 40, 48, 49, 50, 59, 62,
    80, 85, 92, 104, 105, 125, 127, 137, 139,
    142, 155, 159, 160
Lechusa Road, 142, 148
Liebre Mountain, 99, 158
Little Sycamore Canyon Road, 155, 161,
    171
lizards, 42, *43*, 43
Lois Ewen Overlook, 86-87, 91, 94, 99
Lone Oak, 44, 45, *49*, 50, 55
Loop Trail, 195-196
Los Angeles, 10, 27, 34, 35, 39, 44, *45*, 48,
    49, 51, 52, 68, 84, 87, 88, 90, 91, 92, 94,
    99, 100, 175
Los Angeles County, 42
Los Angeles River, 34
Los Angeles Times, 16, 90
Los Padres National Forest, 87
Lower Betty Rogers Trail, 54, 59
Lucky Ranch, 155, 156
lupine, 133

M*A*S*H, 114, 119
MacKaye, Benton, 16
Malibou Lake, 118, 124
Malibu, 12, 34, 35, 81, 88, 90, 95, 117, 118,
    144, 153, 155, 171, 191

Malibu Canyon, 34, 100, 104, 114, *115*, 120, 121
Malibu Canyon Road, 96, 102, 108, 109, 112, 114, *115*, 115, 121
Malibu Coast Fault, 35
Malibu Creek, 24, *108*, 108, 109-110, 111, 118, 121, 122
Malibu Creek State Park, 14, 17, 92, 96, 97, 99, 103, 108, 109, 112, 114, 115, 121, 124
Malibu Creek State Park Campground, 23, 24
Malibu Golf Club, *143*, 144, 145, 150, 151, 156
Malibu Lagoon, 93, 99
Mandeville Canyon, 34
manzanita, 39, *40*, 59, 62, 70, 78, 81, 95, 96, 97, 98, 103, 111, 116, 159, 164, 175, 176
Marvin Braude Mullholland Gateway Park, 45
McAuley, Milt, 16, 120
McAuley Peak, 16, *114*, 120
Mediterranean, The, 37
Mediterranean climate, 37
Mesa Peak, 120, 121
Mesa Peak Motorway, 37, 91, 114-122
Mildas Drive, 92
Miller, Ray, 192
Mishe Mokwa Trail, 170, 172, *174*, 175, 176-178, 187
monkey flower, 10, 39, 125, 128, 142, 160, 172
Monte Nido, *101*, 105, 106, 108, 111, *112*, 112
Morrison, Jim, 117
moss, 41, 72, 78, 80, 81, 82, 83, 90, 92, 97, 101, 103, 104, 106, 108, 112, 121, 133, 137
moths, 72
Mount Abel, 105, 112
Mount Allen. See Sandstone Peak
Mount Baldy, 4, 31, 48, 90, 94, 100, 159, 164
Mount Laguna, 52
Mount Pinos, 103, 112, 118, 157, 164, 166
Mount Rushmore, 176
Mount San Gorgonio, 48, 87, 156
Mount San Jacinto, 52, 94, 100, 166
Mount Triunfo, 157, 158, 160, 163, 166, 167, 172
mountain bikers, 16, 21, 23, 27, 28, 29, 31, 33, 47, 51, 60, 65, 66, 126, 142, 146, 148, 152, 153, 160, 185, 190, 191, 194, 196
mountain lions, 4, 10, *32*, 32, 42, 142
Mountains Restoration Trust, 87
mudstone, 37
Mugu Peak, 195, 196
Muir, John, 5

Mulholland Drive, 86, 91, 96, 102
Mulholland Highway, 23, 86, 91, 96, 102, 118, 124, 138, 142, 144, 148, 150, 152, 154-155, 156, 157, 160, 161, 165, 171, 191
Mulholland Road, 46
Murphy Ranch, 55, 58, 59
Murphy Stairs, 43, 52, 55, 59
Musch Meadows, 41, *62*, 62, *69*, 70
Musch Trail, 15, 16, 60-63, 64, 69
Musch Trail Camp, 23, 24, 62, *63*
mushrooms, 10, 41, 90
mustard, 41, 133

National Park Service, 16, 17, 18, 19, 23, 25, 26, 154, 157, 172, 181
Nelson, Susan, 14
Newbury Park, 178
Newton Canyon, 123, 127-128, 130-134, 136, 137, 138
Newton Canyon Falls, 135, 136-137, 138
Newton Canyon Trail, 15, 16, 130-134
Newton Motorway, 127
Newton Stream, 136
non-native species, 33, 37, 41, 133, 195
North American Plate, 35
Northridge Earthquake, 78, 82

oak woodland plant community, 37, 40, 68, 70, 76, *79*, 79, 80, 81, 96, 103, 108, 111, 125, 128, 135, 138, 139, *141*, 143-144
oak trees, 17, 40, 41, 50, 59, 63, 64, *71*, 71, 72, 74, *75*, 75, 78, 80, 82, 83, 92, 111, 116, 118, 121, 122, 123, 126, 128, 130, *131*, 132, 133, 134, *138*, 177, 183, 184, 186, 198
oak savannah plant community, 37, 41, 58
Ojai, 88
Oklahoma, 44
Old Boney Trail, 182, 183, 185, 198
Old Topanga Canyon, 74, 76
Old Topanga Canyon Road, 76, 79, 86, 89, 91, 96, 102
Old Topanga Canyon Stream, 24, 79
Old Topanga Fire, 81
Orange County, 90, 94, 95, 100
Overlook Fire Road, 186, 187, 192, *194*, 194-195, 198
owls, 42
Oxnard Plain, 34, *170*, 174, 175, 178, 180, 181

Pacific Coast Highway, 29, 34, 61, 69, 74, 79, 86, 91, 96, 102, 109, 115, 124, 131, 132, 135, 142, 148, 154, 161, 171, 187, 191-192, 193, 198, 199
Pacific Crest Trail, 11, 14

Pacific Ocean, 11, 17, 24, 34, 44, 48, 51, 52, 54, *61*, 64, 66, 78, 83, 85, *86*, 90, *91*, 92, 93, 95, 99, 101, 102, *114*, 114, 116, 118, 120, 121, *124*, 124, 131, *133*, 133, 135, 137, 142, 151, 153, 155, 157-158, 160, 163, 164, 165, 167, 169, 170, 173, 174, 175, *179*, *180*, 180, 181, 187, *190*, 190, 191, *192*, 192, 193
Pacific Palisades, 14, 51, 52
Pacific Plate, 35
Pacific tree frog, 43
Palisades Fire, 64, 66
Palisades Highlands, 64, 66
Palm Springs, 166
palm trees, 41
Palos Verdes Peninsula, 44, 48, 52, 64, 67, 85, 88, 90, 93, 99, 120, 124, 125, 133, 164-165
Peak 2049, 16, 120
pepper trees, 41
pine trees, 41, 150
Pine Mountain, 105, 174
pinnacles, *35*, 36, 93, 95, 98, 116, 117, 174, 175, 178, 180
Piuma Ridge, 105, 122
Piuma Ridge Trail, 16, 91, 108-113
Piuma Road, 20, 96, 101, 102, 103, 105, 106, 108, 109, 110-111, 112, 113, 115
Planet of the Apes, 114, 118
Plants, William R., 175
Point Dume, 120, 121, 146
Point Mugu, 10, 14, *190*, 193
Point Mugu State Park, 17, 24, 171, 179, 181, 190, 191
poison oak, *30*, 30, 32, 41, 58, 66, 81, 83
pond lillies, 41
Pop Top, 174, 178
Potrero Valley, 182
prairie falcon, 42
Prier Road, 62
private property, 14, 17, 24, 25, 93, 97, 98, 127, 132, 133, 141, 142, 143, 146, 156, 157
Puerco Motorway, 120

rabbits, 42, 62, 70
raccoons, 42
Rambla Pacifico, 86, 91
Ramirez Canyon, 131, 133
Ranch Center Fire Road, 185
rattlesnakes, 30, 33, 43, 111
Ray Miller Trail, 15, 16, 190, 192-194
Readmond, Darrell, 14
red shank, 39, 96, 155, 159, 160, *163*, 163, 167, 172, 176, 178, 180
reeds, 37, 41, 50, 58, 121, 133, 135, 138, 139, 155, 196

Redondo Beach, 120
Reseda Boulevard, 45, 46, 51, 65
Reseda Spur Trail, 46
Reyes Peak, 182
rideshare services, 28-29, 44, 45, 79, 102, 115, 123, 153, 161, 170, 191
Rim-Of-The-Valley Trail, 14
riparian woodland plant community, 37, 41, 50, 60, 62, 79, 106, 108, 113, 123, 125, 126, 136, 141
Rivas Canyon, 50, 51
Robinson Road, 71
Rock Garden, The, 93, 95, *96*, *97*, 98, *100*
Rocky Mountains, 35
Rocky Oaks Park, 151
Rogers, Betty, 53
Rogers, Will, 44, 45, 48, 52, 53, 54, 55, 58, 59
Rogers Ridge, *45*, 51, 53
Rogers Ridge Trail, 50, 51
Rogers Road, 44, 47, 48
Rosenthal Vineyards, 132, 134
ruins, 55, 58-59, 79, 81, 93, 99, *110*, 111, 116, 138, 151
Rustic Canyon, 36, 41, 43, 44, 45, 47, 48, *49*, 49, 50, 51, 52, 55, 59, 65
Rustic Canyon Trail, 51, 55
Rustic Stream, 59

Saddle Creek, 104
Saddle Creek Trail, 15, 16, 20, 101-106
Saddle Peak, 37, 48, 90, 92, 93, 94, 95, 96, 99-100, 101, 103-105, *105*, 108, *109*, 112, 120, 121, 122, 126, 173, 180
Saddle Peak Road, 78, 83, 86-87, 89, 91
Saddle Peak Trail, 90-100, 105
Saddle Rock, 133, 138, *144*, 144, 146, 156
sage, 39, 49, 50, 54, 81, 104, 105, 116, 132, 137, 139, 142, 151, 155
sagebrush, 39, 40, 104
San Andreas Fault, 35
San Bernardino Mountains, 35, 48, 94, 100, 166
San Diego County, 52
San Fernando Valley, 34, 44, 48, 60, 64, 65, 78, 82, 83, 86, 96, 97, 98, 99, 101, 119, 120, 124, 164
San Gabriel Mountains, 35, 48, 82, 89, 90, 119, 120, 127, 157, 159, 164, 166
San Miguel Island, 181
San Nicholas Island, 94, 100, 133, 155
San Rafael Mountains, 35
sandstone, 35, 36, 37, 62, 63, 64, 65, 66, 70, 72, 80, 85, 92, 93, 98, 103, 104, 116, 117, *118*, 125, 126, 127, 137, 156, 163, 164, 169, 174, 178-179

Sandstone Peak, *31,* 35, 144, 146, 151, 157, 158, 163, 169, 170, 172, 173-174, 175, 176, 187
Sandstone Peak Trail, 172, 175
Santa Ana Mountains, 90, 94, 100
Santa Barbara, 35, 87, 174, 178, 181
Santa Barbara Island, 89, 94, 100, 125, 133, 155
Santa Clara River Valley, 181
Santa Cruz Island, 157, 181, 194
Santa Monica, 44, *51,* 51, 52, 78, 92, 120
Santa Monica Bay, 52, 90, *91,* 92
Santa Monica Mountain Range, 10, 12, 13, 14, 16, 17, 18, 31, 32, 34, 35, 36, 37, 39, 41, 42, 44, 45, 48, 60, 64, 68, 78, 90, 95, 99, 114, 121, 127, 132, 138, 148, 153, 163, 164, 169, 170, 173, 174, 178, 180, 181, 187
Santa Monica Mountains Conservancy, 13, 14, 16
Santa Monica Mountains National Recreation Area, 12, 13, 14, 17, 21, 25, 97
Santa Monica Mountains Task Force, 15, 16
Santa Monica Mountains Trails Council, 14, 15, 16, *25,* 25
Santa Monica Slate, 36
Santa Rosa Island, 157, 181, 194
Santa Suzanna Mountains, 35, 78, 82, 118
Santa Ynez Canyon, 64, 66
Santa Ynez Mountains, 35, 87, 174, 181
Santa Ynez Trail, 64
Satwiwa Cultural Center, 178
Scenic Trail, 187, 198
Schueren Road, 86-87, 91, 92, 103
Schwarzenegger, Arnold, 12, 142, 145
scrub oak, 38, 39, 48, 105, 155
Sepulveda Canyon, 34
Serrano Canyon Trail, 186, 198
Serrano Valley, 182, 186, *193,* 193, 194, 198
Sespe Condor Refuge, 146
Sespe Formation, *35,* 36-37, 80, 85, 116, 125, 156, 164
shale, 128
Sherman, Alan, 62
Sierra Club, 15, 16, 25
Sierra Madre Mountains, 35
siltstone, 36
Silver Moccasin Trail, 14
Simi Hills, 103, 167
Simi Valley, 104, 105
Skull Rocks, The, 176, 187
Snakebite Ridge Road, 130, 132, 133
Solstice Canyon, 24, 34, 123-127, *124, 126, 127*
Solstice Creek, 125
southern Pacific rattlesnake, 43

Sphinx, The, 176
Split Rock (on Chamberlain Trail). See Chamberlain Rock
Split Rock (on Mishe Mokwa Trail), 172, 176, *177,* 177, 187
Springs Fire, 17, 183, 196, 198
squirrels, 40, 42, 62, 70
Studio City, 14
Stunt Road, 20, 86-87, 91, 92, 96, 97, 101, 102-103
succulents, 32, 38, 40
Sullivan Fire Road, 59
Summit Trail, 71
Sunset Boulevard, 45
Swift, Jill, 14
switchbacks, 33, 81-82, 83, 89, 96-97, 104, 105-106, 112, 119, 121, 125, 126, 128, 144, 153, 155-156, 163, 167, 172-173, 182, 183
sycamore trees, 12, 17, 41, 66, 71, *72, 102,* 106, 123, 126, 176, 184, 186, 195, 198
Sycamore Canyon Campground, 171, 186, 187, 191-192, 198-199
Sycamore Canyon Falls, 178
sycamore savannah plant community, 37, 41, 58, 169, 184

Tapia Park, 24, 106, 109, 110, 115
Tarzan films, 114, 118
Taylor, James, 38
Tehachapi Mountains, 104
Temescal Canyon, 48, 49
Temescal Gateway Park, 47
Temescal Peak, 47, 48, 173
Temescal Ridge, 49
Temescal Ridge Fire Road, 48
Temescal Ridge Trail, 46, 47
thistle, 41, 111, 112
Thornhill Broome Beach, *192,* 193
Thousand Oaks, 124, 153, 157, 166, 167, 174, 183, 195
Tom Harrison Maps, 22
Tongva people, 104, 117
Topa Topa Peak, 35, 88, 174
Topanga, 22, 61, 62, 64, 69, 79, 81
Topanga Canyon, 34, 36, 48, 60, 61, 62, 63, 68, 71, 74, 75, 78, 80, 82, 83, 121
Topanga Canyon Boulevard, 46, 61, 69, 71, 72, 73, 74, 75, 79, 86, 91
Topanga Elementary School, 74, 75, 76
Topanga Formations, 37
Topanga Lookout, 82, 84, 86, 87-88
Topanga State Park, 12, 42, 45, 46, 53, 60, 61, 64, 69, 79
Topanga Stream, 24, 75
Topanga Tower Motorway, 87

toyon, 39, *40*, 40, 48, 50, 59, 62, 71, 80, 92, 98, 103, 104, 127, 137, 163
Trail Magic Adventures, 25
Trancas Canyon, 34, 36, 40, 141-147, 149, *151*, 151, 156
Trancas Canyon Road, 42, 144, 145
Trancas Stream, 24, 143
Transverse Ranges, 35
tree tunnels, 38, *87*, 96, 112, 113
Tri Peaks, 174, 178, 179, 187
Trippet, Cora, 61
Trippet, Oscar A., 61
Trippet, Oscar Jr., 61
Trippet Creek, 71
Trippet Ranch, 24, 42, 45, 47, *60*, 60, 61, 68, 69, 70, 75
Trippet Ranch Visitors Center, 69
Trippet Ranch Parking Lot, 64, 70
Triunfo Lookout, 164, *166*, 166, 172, 173
Triunfo Pass, *11*, 21, 23, 32
Triunfo Pass Parking Lot, 32, 162-163, 164, 171, 177
Triunfo Pass Satellite Earth Station, 157, 165
Turritella fossils, 37, 119, *121*
Twilight Zone, The, 97
Two Foxes Trail, 185

Udell Gorge, 118
Uinta Mountains, 35
United States Geological Survey, 17
Upper Betty Rogers Trail, 53, 54, 59
Upper Newton Canyon. See Newton Canyon
Upper Solstice Canyon. See Solstice Canyon
Upper Zuma Canyon Trail, 135-140
Upper Zuma Falls, 135, 136, 138-139, *139*
Utah, 93, 95, 98, 169

Valmar Road, 86, 91, 96, 102
Vaqueros Formation, 37
Ventura, 87
Ventura County, 10, 88, 104, 105, 155, 158-159, 161, 166, 171, 174, 182
vineyards, 130, *132*, 132, 133

Warner Center, 99
water tanks, 66, 76, 90, 92, 116, 118, 120, 121, 175
waterfalls, 11, 12, 27, 104, *106*, 125, 135, 136-137, 138-139, *139*, 178, 195, 196
Webster, Ron, 16, 192
Weider, Betty, 145
Wendy Drive, 183

western mountain mahogany, 80, 105, 139, 142, 155
Westlake Boulevard, 154, 161, 171
Westlake Village, 157, 158, 195
Westside, The, 49, 60, 86
wild asparagus, 58
wild grape, 41, 106
wildflowers, 48, 105, 123, 134, 148, 190, 192, 196
Will Rogers State Historic Park, 14, 15, 23, 24, 42, 44, 45, 47, 48, 51, 53, 55, 59, 65
willow trees, 41
Wilshire Boulevard, 49
Wood Canyon, 196, *198*, 198
Wood Canyon Fire Road, 185, 195, 196
Wood Canyon Vista Trail, 16, 185, 186, 195, 196-198, *198*
Woolsey Fire, 17, 118, 123, 130, 135, 138, 141, 143, 150, 158, 160, 166, 167, 177, 183
World War II, 58

Yellow Hill Fire Road Trail, 164, 166
Yerba Buena Road, 23, 153, 155, 156-157, 158, *159*, 159, 162, 163, 166, 167, 169, 171, 172, 177
Youth Conservation Corps, 104, 132
yucca, 10, 29, 32, 39, 49, 54, 81, 105, 114, 116, 119, 123, 127, 137, 139, 143, 144, 148, 149, 151, 155, 160, 164, 172, 184, 190, 192, 195

Zuma Beach, *142*, 146, *151*, 151, 156, 157
Zuma Canyon, 34, 121, 128, 132, 134, 135-139, *136*, 142, 146, 147
Zuma Creek, 24, *137*, 137, 139
Zuma Ridge, 144, 145, 146, 156
Zuma Ridge Motorway, 15, 136, 139, 140, 144, 145-146

101 freeway, 17, 34, 61, 69, 74, 79, 86, 91, 96, 102, 109, 115, 124, 131, 136, 148, 154, 161, 171, 174
127 Hours, 31

20th Century Fox, 114
23 freeway, 174

405 freeway, 34, 45

92 Trail, 68, 70, 71
92 Spur Trail, 70

# ABOUT THE AUTHORS

Husband and wife Doug and Caroline Chamberlin are avid hikers who live in Southern California, but who have enjoyed hiking in areas as diverse as the Colorado Rockies, the Grand Canyon, the Italian Dolomites and the Scottish Highlands. They fell in love with the Backbone Trail when they first hiked it in 2011. Since then they have trekked the entire trail several times, and have the retired pairs of pulverized hiking boots to prove it.

Doug is a professional writer and semi-professional graphic designer who got the hiking bug as a child on family trips in the Appalachians. He is mostly known for his work in screenplays, including co-authoring the script for *Toy Story II* as well as numerous television shows. He writes humorous non-fiction books and blogs a humor column.

Caroline became a convert to hiking after she met Doug, when she moved to Southern California and fell in love with the mountains. She is a Certified Public Accountant of over twenty years.

You can contact Doug and Caroline at **info@backbonetrailguide.com**.

*The authors, out standing in their field*

Made in the USA
Middletown, DE
05 October 2022